Lecture Notes in Computer Science 12060

More information about this series at http://www.springer.com/series/8183

Marina L. Gavrilova · C. J. Kenneth Tan ·
Alexei Sourin (Eds.)

Transactions on Computational Science XXXVI

Special Issue on Cyberworlds and Cybersecurity

Springer

Editors-in-Chief
Marina L. Gavrilova
University of Calgary
Calgary, AB, Canada

C. J. Kenneth Tan
Sardina Systems OÜ
Tallinn, Estonia

Guest Editor
Alexei Sourin
Nanyang Technological University
Singapore, Singapore

ISSN 0302-9743   ISSN 1611-3349   (electronic)
Lecture Notes in Computer Science
ISSN 1866-4733   ISSN 1866-4741   (electronic)
Transactions on Computational Science
ISBN 978-3-662-61363-4   ISBN 978-3-662-61364-1   (eBook)
https://doi.org/10.1007/978-3-662-61364-1

This Springer imprint is published by the registered company Springer-Verlag GmbH, DE
part of Springer Nature
The registered company address is: Heidelberger Platz 3, 14197 Berlin, Germany

LNCS Transactions on Computational Science

Computational science, an emerging and increasingly vital field, is now widely recognized as an integral part of scientific and technical investigations, affecting researchers and practitioners in areas ranging from aerospace and automotive research to biochemistry, electronics, geosciences, mathematics, and physics. Computer systems research and the exploitation of applied research naturally complement each other. The increased complexity of many challenges in computational science demands the use of supercomputing, parallel processing, sophisticated algorithms, and advanced system software and architecture. It is therefore invaluable to have input by systems research experts in applied computational science research.

Transactions on Computational Science focuses on original high-quality research in the realm of computational science in parallel and distributed environments, also encompassing the underlying theoretical foundations and the applications of large-scale computation.

The journal offers practitioners and researchers the opportunity to share computational techniques and solutions in this area, to identify new issues, and to shape future directions for research, and it enables industrial users to apply leading-edge, large-scale, high-performance computational methods.

In addition to addressing various research and application issues, the journal aims to present material that is validated – crucial to the application and advancement of the research conducted in academic and industrial settings. In this spirit, the journal focuses on publications that present results and computational techniques that are verifiable.

Scope

The scope of the journal includes, but is not limited to, the following computational methods and applications:

- Aeronautics and Aerospace
- Astrophysics
- Big Data Analytics
- Bioinformatics
- Biometric Technologies
- Climate and Weather Modeling
- Communication and Data Networks
- Compilers and Operating Systems
- Computer Graphics
- Computational Biology
- Computational Chemistry
- Computational Finance and Econometrics
- Computational Fluid Dynamics

- Computational Geometry
- Computational Number Theory
- Data Representation and Storage
- Data Mining and Data Warehousing
- Information and Online Security
- Grid Computing
- Hardware/Software Co-design
- High-Performance Computing
- Image and Video Processing
- Information Systems
- Information Retrieval
- Modeling and Simulations
- Mobile Computing
- Numerical and Scientific Computing
- Parallel and Distributed Computing
- Robotics and Navigation
- Supercomputing
- System-on-Chip Design and Engineering
- Virtual Reality and Cyberworlds
- Visualization

Editorial

The *Transactions on Computational Science* journal is part of the Springer series *Lecture Notes in Computer Science*, and is devoted to the gamut of computational science issues, from theoretical aspects to application-dependent studies and the validation of emerging technologies.

The journal focuses on original high-quality research in the realm of computational science in parallel and distributed environments, encompassing the theoretical foundations and the applications of large-scale computations and massive data processing. Practitioners and researchers share computational techniques and solutions in the aforementioned areas, identify new issues, and shape future directions for research, as well as enable industrial users to apply the presented techniques.

The current volume is devoted to state-of-the-art approaches in the domain of Cyberworlds and Cybersecurity. This issue contains two parts: the first part comprises of four papers selected following the annual 2018 International Conference on Cyberworlds and is edited by Alexei Sourin, while the second part contains two manuscripts accepted from an open CFP, covering the topics of fast 3D segmentation using geometric surface features and nature-inspired optimization for face recognition. All the accepted papers have been peer-reviewed.

We would like to extend our sincere appreciation to the Special Issue Guest Editor, Alexei Sourin, for his continuous dedication and insights in preparing this special issue. We would also like to thank all of the authors for submitting their papers to the journal and the associate editors and referees for their valuable work.

It is our hope that the collection of papers presented in this special issue will be a valuable resource for *Transactions on Computational Science* readers and will stimulate further research into the vibrant area of computational science applications.

March 2020   Marina L. Gavrilova
                                                                     C. J. Kenneth Tan

Guest Editor Preface

Created intentionally or spontaneously, cyberworlds are information spaces and communities that use computer technologies to augment the way we interact, participate in business, and receive information throughout the world. Cyberworlds have ever-growing impact on our lives and the evolution of the world economy. Examples include social networking services, 3D shared virtual communities, and massively multiplayer online role-playing games. Problems of cyberworlds were discussed at the annual 2018 International Conference on Cyberworlds which was held in Singapore on during October 3–5, 2018. Several papers were invited to contribute with extended and revised articles to this issue of the *Transactions on Computational Science*.

The paper "Orion : A Generic Model and Tool for Data Mining" by Cédric Buche, Cindy Even, and Julien Soler focuses on the design of autonomous behaviors based on humans behavior observations. In this context, the contribution of the Orion model is to gather and to take advantage of two approaches: data mining techniques (to extract knowledge from the human) and behavior models (to control the autonomous behaviors).

The paper "Environment Estimation for Glossy Reflections in Mixed Reality Applications Using a Neural Network" by Tobias Schwandt, Christian Kunert, and Wolfgang Broll considers a topic of environmental textures. The authors propose an image stream stitching approach combined with a neural network to create plausible and high-quality environment textures that may be used for image-based lighting within mixed-reality environments.

In the paper "Distance Measurements of CAD Models in Boundary Representation" by Ulrich Krispel, Dieter W. Fellner, and Torsten Ullrich the problem of calculating the distance between two polygonal objects in real-world scenarios is addressed. The authors contribute a publicly available benchmark to compare distance calculation algorithm, and they also investigated and evaluated a grid-based distance measurement algorithm.

Finally, the fourth paper "An Immersive Virtual Environment for Visualization of Complex and/or Infinitely Distant Territory" by Atsushi Miyazawa, Masanori Nakayama, and Issei Fujishiro presents an immersive virtual environment that allows the user to set environmental limits beyond three-dimensional Euclidean space. This is achieved by setting the limits to n-dimensional complex projective space, an element of both complex and infinitely distant domain can be naturally visualized as a recognizable form in the Euclidean 3-space.

The organizers of the conference are very grateful to Prof. Marina Gavrilova, Editor-in-Chief of the *Transactions on Computational Science*, for her continuing support and assistance. We also thank the reviewers for their invaluable advice that helped to improve the papers.

March 2020

Alexei Sourin

Contents

ORION: A Generic Model and Tool for Data Mining

Cédric Buche[1](✉), Cindy Even[1,2], and Julien Soler[2]

[1] Lab-STICC CNRS UMR 6285, ENIB, 25 rue Claude Chappe,
29280 Plouzané, France
{buche,even}@enib.fr
[2] Virtualys, 41 rue Yves Collet, 29200 Brest, France
julien.soler@virtualys.com

Abstract. This paper focuses on the design of autonomous behaviors based on humans behaviors observation. In this context, the contribution of the ORION model is to gather and to take advantage of two approaches: data mining techniques (to extract knowledge from the human) and behavior models (to control the autonomous behaviors). In this paper, the ORION model is described by UML diagrams. More than a model, ORION is an operational tool allowing to represent, transform, visualize and predict data; it also integrates operational standard behavioral models. ORION is illustrated to control a bot in the game Unreal Tournament. Thanks to ORION, we can collect data of low level behaviors through three scenarios performed by human players: movement, long range aiming and close combat. We can easily transform the data and use some data mining techniques to learn behaviors from human players observation. ORION allows us to build a complete behavior using an extension of a Behavior Tree integrating *ad hoc* features in order to manage aspects of behavior that we have not been able to learn automatically.

1 Introduction

This research focuses on the implementation of behaviors to provide skillful and believable Non-Player Characters (NPCs). In this context, imitation learning is a very promising way to build such behaviors [6]. Imitation learning consist in extracting knowledge from data produced by human players in order to be able to reproduce their behaviors. Data mining offers a large range of tools that can be very useful to extract important knowledge.

Datasets obtained from human players can be of various types and forms and are used to learn behaviors. They can be time-series or not, contain quantitative and qualitative attributes, be abstract or have strong semantics. No model covers all these cases. The data semantics is generally not taken into account in different data processing and representation tools such as ELKI [1] WEKA [14], GGobi [23] or Orange Canvas [8]. Even if these tools allow to visualize data through scatter plots, histograms and other diagrams, in order to perform an

M. L. Gavrilova et al. (Eds.): Trans. on Comput. Sci. XXXVI, LNCS 12060, pp. 1–25, 2020.
https://doi.org/10.1007/978-3-662-61364-1_1

efficient analysis, it is essential to visualize the data according to their semantics, means make the connection between data and their meaning. The problem is well identified by Vondrick et al. [30]. The authors present studies on object detection in an image using feature extraction techniques. Once feature extraction is done, the data becomes abstract. Investigating why a particular data is misclassified, for example, becomes extremely complex. In order to maintain understandability of data throughout the analysis process, the authors propose an image reconstruction from the extracted features. This is indeed a major problem that must be addressed. Visualization and transformations applied to the data should be consistent with their semantics. For example, given that the data represents two or three dimensional vectors, it is relevant to calculate an angle between two of them. The semantics also help the user to choose a more appropriate distance between data. It is certainly better suited to compare two RGB colors using a DeltaE distance rather than an Euclidean distance.

Existing tools suffer from limited data visualization possibility, providing numeric and categorical data only and never make the link with behavioral model. ORION tackles those issues. The ORION model proposes a generic approach to represent datasets. It also offers the possibility to perform some transformations on these datasets and to visualize them. In addition, the ORION model makes the link between data and behavioral models.

In Sect. 2, we present the ORION model. In this section, we present the workflow (Sect. 2.1) and we detail the model (Sect. 2.2). The model is described using UML diagrams as a generic data mining model (Sect. 2.2) and a behavioral model linked to the data mining model (Sect. 2.2). More than a model, ORION is also available as an operational tool. ORION is then a full software; it can be used as a library if the user need to define new types, algorithms or behaviours. In Sect. 3, we illustrate the use of the ORION tool to control a virtual player in the game Unreal Tournament 3. Section 4 provides conclusion and future works.

A preliminary version of this work has been reported in [7]. This paper is an extended version, including in-depth state of the art, technical details and examples to illustrate the proposed models.

2   ORION

2.1   ORION Workflow

We offer with ORION a complete work-flow divided into two parts: a structural and a behavioral part (see Fig. 1).

Taking the example of a 3D video game, data are usually strongly associated to concrete concepts such as position, speed, orientations, hit points, etc. Other examples are given in the UCI[1] database where one can find datasets provided as CSV files, representing images of handwritten characters, chess positions, musical notes, etc. In ORION, the first task for the analyst is to add semantics to the data. Then, he can transform and visualize them according to their semantics

---

[1] https://archive.ics.uci.edu/ml/datasets.html.

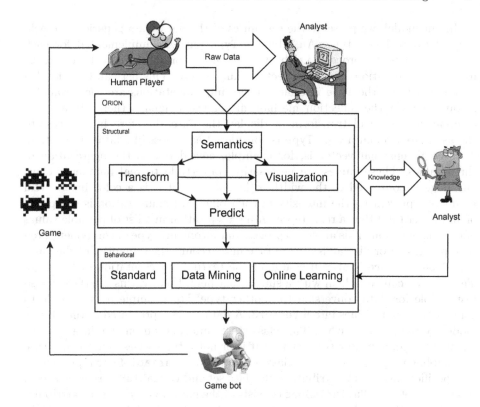

Fig. 1. ORION workflow

which ease the exploratory analysis. The analyst may find in the data, invariants between players for example. He can also train predictive models with these data. After this analysis step, the analyst uses the behavioral model to build a behavior using information extracted from his preliminary analysis and his predictive models. ORION's behavior model is based on a paradigm commonly used in the video game's industry. Therefore, the analyst has access to traditional tools used for IA (behavioral block). He can easily implement the knowledge identified in the data (to reproduce the invariant observed between the players for example). He can also use the predictive models that he built to reproduce a part of the behavior (like the aiming of an enemy for example).

2.2   ORION Model

ORION Structural Model

Data Representation. Traditional data processing and representation tools do not take data semantics into account [1,8,14,22]. Without semantics, data become abstract after the feature extraction. Visualizing data according to their semantics facilitates their analysis [30].

In our model, we preserve the semantics of the data via a generic approach: each attribute has a type. An interface `Type` must be implemented for each type of data to be processed. The ORION model provides basic types (reals, integers, enumerations) but also vectors, images, etc. This list can be extended by implementing the `Type` interface. These implementations are then available in our tool via the reflection mechanism[2] of the language (here, Java). The components that will transform or display the data therefore have access to the associated semantics. A `Type` is considered as a possible data aggregation. A three-dimensional vector is, for example, considered as the aggregation of three values. These values are necessarily convertible to real numbers. Thus, an image is composed of the width × height (× 3 if it is a color image) real numbers representing the intensity of each pixel. An enumeration is composed of an integer (and thus a real) representing its position in a list of possible values of the enumeration. This allows to process data from any type of component even if no processing or display is able to take into account the semantics of the data. The `Type` interface is also in charge of the textual representation of the data. Thus, a user can edit them with a simple text field. Finally, this interface is also responsible for data conversion to another type. For example it is possible to convert three real values into a `Vector3`. A data set is represented in the ORION model by the `Dataset` class. This class stores information on the data set, like for instance its name, or the fact that it is or not a time series, and in that case its sampling frequency, etc. This class is composed of `VariableSpec` representing the specification of each attribute in the data set and `VariableSet` representing a given data set. The `VariableSpec` consists of the name and type of the attribute, and whether the attribute is an input or an output. The `VariableSet` consist of a list containing their values referenced by the `VariableSpec` of the corresponding attribute. Appendix 1 shows the UML class diagram that implements our data representation. The `Dataset` class is also composed of an attribute `summaryData` of type `VariableSet`. This attribute is mainly used to store the result of vector quantization for example. Also, attributes can be converted from one type to another.

To illustrate data conversion, we present in Fig. 2 conversion process of the Optitracks database available on the UCI database. This dataset represents images of handwritten digits. The original image was made up of two colors and had a resolution of 32 × 32 pixels. Each image was cut into blocks of 4x4 pixels. The dataset consists of 65 attributes. The first 64 attributes represent the number of black pixels in each block of 4 pixels by 4. The 65th represents the class of the data (a digit). After importing the csv le into the tool, the process is made of five steps: (a) selection of the attributes to be converted, (b) specification of the parameters of `VariableSpec` into which the data will be converted (the name of the attribute, input or output), (c) type selection and optionally (d) configure the parameters of this type (here it is the image type, so we need to specify its dimensions). Step (e) shows the result of the conversion of the first

---

[2] Reflection provides information about the class to which an object belongs and also the methods of that class which can be executed by using the object.

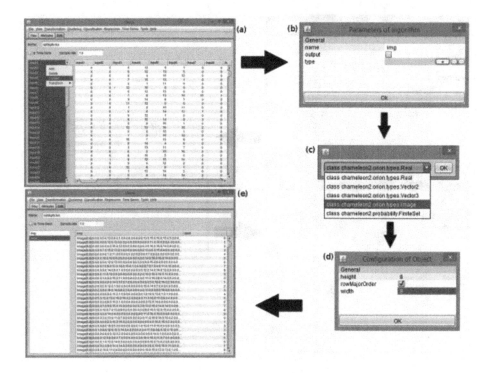

Fig. 2. Type conversion process for the optitrack database in ORION.

64 attributes in an attribute (which was the type Real) to a type Image and the 65th `FiniteSet` type.

Data Transformation. In order to perform data transformations, the generic `DataTransform` interface must be implemented. The latter simply has access to the `Dataset` and a list of `VariableSpec` to process and must implement the `DataTransform.transform()` method. Allowed transformations can vary significantly. They generally involve attributes modification, addition of attributes or `summaryData`, etc. The downside of this simplicity is that the GUI for manipulating this model has no information on the type of transformation performed by the `DataTransform`. In order to overcome this problem, various sub-interfaces are proposed like `ClusteringTransform`, `DiscretizerTransform`, `VectorQuantization`, etc. These interfaces are only present to classify data transformation methods into several categories. An algorithm can also implement several of these interfaces. Appendix 2 shows the UML class diagram implementing the transformation model of ORION. Data transformation algorithms implemented in ORION include Principal Component Analysis (PCA), Kernel PCA [21], Classical Multidimensional Scaling [28], Isomap [26], Sammon projection [20], and many clustering algorithms (KMeans, DBSCAN [9], OPTICS [2], Gaussian Mixture Model, Growing neural Gas Networks [12], etc). This list can be extended by implementing the `DataTransform` interface.

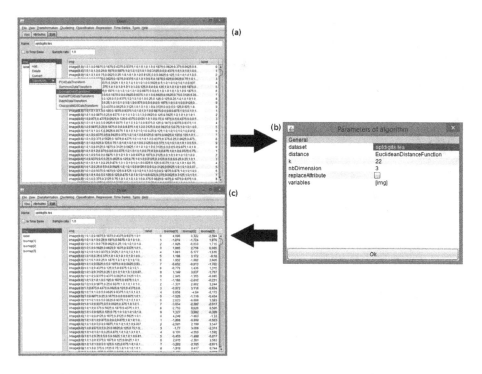

Fig. 3. Transformation process for the optitrack database (Isomap).

Using the example of UCI's Optitracks database, original images used to build the dataset are not available because each image is represented by the sums of black pixels in a block of 4 by 4 pixels. However, it is possible to reconstruct 8×8 grayscale images with this data. To do this, we need to normalize each value between 0 and 1 (1 being white and 0 black). ORION provides a class BatchDataTransform to apply an expression to each value of the data set. In addition, it may be useful to produce multidimensional scaling on data. Figure 3 illustrates the multidimensional scaling execution process carried out by the Isomap algorithm through our tool. Data normalization was previously carried out via a similar process.

Data Visualization. In the ORION model, every element in charge of data visualization implements the DataViewer interface. This interface is configurable by the enumeration DisplayRange to define the data to be represented (the current datum, the current selection or all data). Two display types are present: the 2D and 3D rendering interfaces implements the GraphicalViewer interface while the various components displaying data, attributes by attributes, implement the AttributeViewer interface. The GraphicalViewer interface is implemented by the classes Viewer2D and Viewer3D. This interface is not intended to be extended by other classes. A GraphicalViewer is composed of a Renderer. The Renderer interface provides methods (draw2D and draw3D) to draw a datum

on the display panel and a method to provide a dialog panel to configure the
`Renderer`. For example, the implementation of `PointRenderer` allows to represent data as points. The position of the point can be defined by an attribute
of type `Vector2` or `Vector3` or three `Real` attributes. Similarly, its size, shape
or color can be specified as fixed, varying with an attribute value or with its
position in the time series. This list can be extended by implementing the
`Renderer` interface. Figure 4a and b show the display of UCI's Optitracks dataset
as well as the display of NAO robot's skeleton taken from a goalkeeper agent in
the context of the RoboCup 3D Soccer Simulation Competition. The interface
`AttributeViewer` consists of a list of attributes to display. ORION implementations of `AttributeRenderer` include Radar and parallel plots, histograms, line
charts, etc. This list can be extended by implementing the `AttributeViewer`
interface. Appendix 3 shows the UML class diagram that implements the visualization model of ORION.

Prediction. Prediction is an important task of our data-based learning approach.
In order to achieve prediction and therefore improve learning functions, the
ORION model offers a level of abstraction for all methods of classification and
regression. Appendix 4 illustrates this model. The `FunctionLearner` interface
represents a regression or classification algorithm. This interface is composed
of the explanatory variables and the dependent variables. It has the methods
`predict` and `train` that are respectively used to predict the dependent variable
based on explanatory variables and to train the algorithm based on sample data.
The `Classifier` and `Regressor` interfaces extend `FunctionLearner`.

Several interfaces extend `Classifier` to separate different characteristics of
these algorithms. Indeed, many classification algorithms are binary classifiers
(although most classification methods being initially binary have seen multiclass
extensions proposed like for SVM and AdaBoost). In addition, many algorithms
do not perform a peremptory classification but instead return a probability of
belonging to each class. Interfaces `BinaryClassifier`, `ConfidenceClassifier`
and `BinaryConfidenceClassifier`, therefore express these features. Regression and classification algorithms implemented in ORION include KNN, Quinlan
C4.5 [18], SVM [29], Feed Forward Neural Networks [4], Hidden Markov Model
[31], Naive Bayesian Networks [19], AdaBoost [11], etc. This list can be extended
by implementing the `Classification` or `Regressor` interface.

ORION Behavioral Model. ORION's behavioral model is an extension of
behavior trees (BT) [15]. This section presents the concepts underlying this
model and the extensions made in ORION. Finite State Machine (FSM) and
BT are the architectures the most used in the game industry. FSM has many
shortcomings over BT. Transitions from one state to another must be explicit,
resulting in excessive complexity when the number of states is large. As a state
machine is not a Turing machine, some behaviors are impossible to describe
with FSM. FSMs are event oriented (events are needed to cause state changes)
as opposed to goal oriented, which makes it difficult to implement goals with

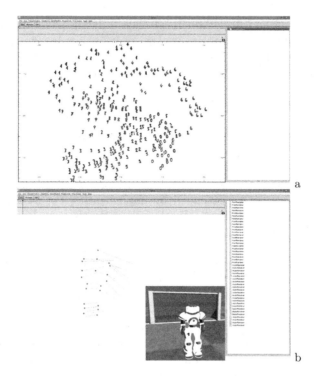

Fig. 4. UCI Optitracks dataset visualization (a). NAO Skeleton dataset visualization with ORION (bones positions and speed). The picture at the bottom right, that is not part of ORION GUI, illustrate NAO within the simulation (b)

priority management among them for example. BT do not have these problems and for these reasons we believe that BT is better suited for the development of an AI solution in video games.

Standard Behavior Trees. BTs originate from the computer game industry as a powerful tool to model the behavior of NPC, used by video games such as Halo, Bioshock, and Spore. BT have the ability to create complex tasks composed of simple tasks, without worrying how the simple tasks are implemented. BT model is represented by a tree structure where each node of the tree represents a goal and each child of a node represents a sub goal. When the nodes are executed, they return their current status (Success, Fail or Running). In its most simple implementations, at each time step, the tree is traversed from the root node spreading deep into the branches to the leaves. The leaves of the trees are mainly of two types: condition and action. They are used to condition the execution of actions or goals. Action nodes perform operations on either the agent environment or its internal state. Composite (Non Leaf) nodes are used to control the execution of their child nodes. There are mainly two types: Sequence and Selector. A Selector node executes its child nodes until one of them returns Success. It allows to reach a goal by testing several solutions. A Sequence node executes its child nodes until they all return Success. It allows to perform a sequence of actions to achieve a goal. Decorator nodes have only one child node. These nodes are used to add execution control features like *while* or *for* statement equivalent.

Orion Behavior Tree Extensions. We saw how BT is endowed with many qualities to achieve an AI in video games, therefore we decided to extend this model. First, without losing the potential of BT to add *ad hoc* code features, we use as an execution context of BT, a `VariableSet`. This context contains, at each tick, data collected by the agent and allows the different nodes to change this context (add or change data). The data being assigned to a type, each behavior can control its consistency and performs introspection on these data. We propose to extend BT to be able to use the data mining techniques. We presented earlier, the structural model encompassing the various tasks achievable by data mining techniques. Here we present the integration of the structural model into the behavioral model.

The most important functionality we want to add to BT is the possibility to use behaviors that have been learned from a dataset. To do this, we propose to add node types to the BT model. First, we add a particular type of node composed of a `FunctionLearner` to predict an output variable based on input variables. This is an action node that changes the context by adding the predicted value in it. We also add a new type of composite node to choose which child node to run through the prediction of a classifier. Finally, we add an action node composed of a `DataTransformer`, to transform the data. Appendix 5 presents our model of BT.

We want the construction of BT to be possible both online and offline. The offline construction is simply enabled by our implementation of Orion, manually constructing the tree and setting the node parameters (by associating a `FunctionLearner` previously trained with data to a `FunctionBehavior` for example). But online training requires to integrate other mechanisms. Indeed, many additional problems arise when trying to learn online. Most data mining techniques we studied previously are not iterative so it is not possible to start learning at each tick. We must therefore offer a flexible method to trigger a learning algorithm, following an event in the game or after a number of ticks for example. BT's formalism allows to simply achieve this kind of behavior. We decide to add to Orion some nodes dedicated to online learning. These nodes allow to store online data within a dataset and to launch the training over those online datasets. Appendix 6 illustrates the implementation of these features in the Orion model.

The typical use of these units in order to perform online learning is shown in Fig. 5. In this example, the left side of the tree allows the execution of the behavior performed once learning is done and the right part allows directed learning and implement a default behavior if learning has not been achieved. Thus, if learning has not yet been done, the condition node on the left (*Trained?* in the figure) returns `Fail` and the right node is executed. The current context is then stored in a `Dataset`. If the number of data in the `Dataset` is sufficient or an event occurs in the game for example, learning is carried out in several steps. First, the `Dataset` is divided into n sequences by a data processing algorithm (usually a time series clustering algorithm). These sequences belong in this example to two different types. Then, the three `FunctionLearner` nodes associated both with the execution part (the three nodes references *Classifier 1*, *Regression 1* and *Regression 2* in the Fig. 5) and the Train nodes are trained with the `Dataset`. Finally, the execution context is changed to indicate that learning was achieved (Trained = True). The following tick therefore executes the train behavior (left side of the tree). As long as learning is not finished, the default behavior (implemented with *ad hoc* features for example) is executed.

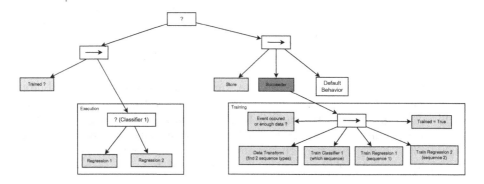

Fig. 5. Typical use of ORION behavior Trees for online learning

3   Application

Different approaches have been explored by the scientific community to design a solution for imitation learning of behaviors. Tencé et al. [24] suggested a probabilistic approach. Connectionists [27] and memory based [5,32] approaches have also been considered. In the video game Black & White (Lionhead Studio, 2001), a decision tree [10] was used to implement a creature of the game. The problem with most of those techniques is that everything is encapsulated in a black box, data from human tracking are not analyzed before entering the imitation learning algorithms. Thus, important data are missed and insignificant data are considered. In order to imitate human behavior, we first need to collect data from human players operating in the game environment and to extract knowledge from these data.

The use of ORION in the design of a Bot for the game Unreal Tournament 3 (UT3) is described in this section. This game is a First Person Shooter (FPS) video game developed by Epic Games and released in 2007. In this game, the player incarnates a character with a variety of futuristic weapons, evolving in an arena with other players. The arena contains items that players can collect. They may be weapons, ammunition, shields, first aid kits etc.

First we describe how by tracking a human player we can collect data (Sect. 3.1) that can be easily visualized and transformed with ORION (Sect. 3.2). We continue by explaining how by predicting the data with ORION we can learn low level behaviors (Sect. 3.3). Next, we show how our behavioral model uses the learned behaviors and how we implement features of the Bot behavior which we have not been able to learn automatically (Sect. 3.4). Finally, we provide results (Sect. 3.5).

An overview of different components of the ORION model used in this application is presented in Fig. 6.

3.1 Human Players Tracking

Raw Data. When observing a player evolving in the game with GameBots, we have access to much information about him and his environment. We choose to collect this unprocessed information in a CSV file. Raw data are processed with ORION as explained later. The information collected is summarized in Table 1.

Table 1. Data collected

Player position
Player orientation (pitch and yaw, roll is always 0)
Player velocity
Current weapon name
Life points
Remaining ammunition
Shoot (is the player shooting?)
Number of enemies visible
Enemy positions (the 3 closest)
Enemy orientations (the 3 closest)
Enemy velocities (the 3 closest)
Number of seen items
Item positions (the 3 closest)
Number of seen navigation points
Navigation point positions (the 3 closest)

Data Collection. In order to create datasets containing only sequences where a player is performing a specific task, we conducted scripted gaming sessions. A human player performs the tasks defined in our scenarios and game traces are recovered. Three scenarios are proposed:

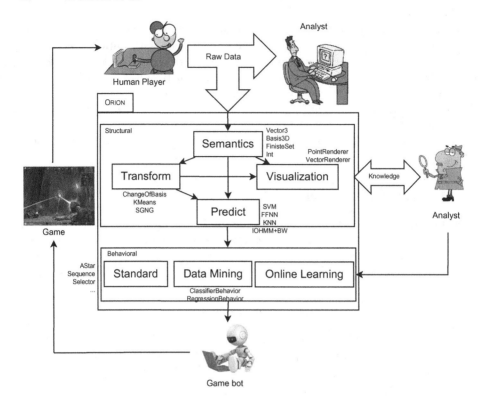

Fig. 6. ORION Workflow for UT3

1. Movement: The player must simply navigate in the environment. No enemy is present. The obtained data will allow us to learn how to move in the environment.
2. Long range aiming: The player cannot move. He has infinite ammunition. He is alone in a large room with a single enemy and only intended to shoot it.
3. Close Combat: The player is in a small room with an enemy and must dodge attacks while firing at the enemy.

3.2   Data Transformation and Visualization

Raw data does not allow to easily perform analysis. As ORION adds semantics to the data, we make use of it to facilitate an exploratory analysis. In UT3, most data are spatial locations of elements in the game (players, items, etc). ORION implements several types associated with locations in three-dimensional space and allows their visualization. Using these features, we are able to rebuild the game play in ORION as shown in Fig. 7.

Spatial location from raw data does not allow us to effectively reproduce the observed behavior. The position of the player, for example, should not be used

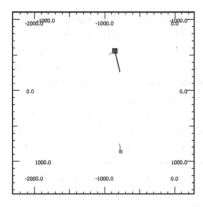

Fig. 7. UT3 game reconstruction in ORION

as input data of a learning algorithm. Doing so, the learned behavior would only be able to reproduce actions at the specific locations where they were observed. Therefore, we need to transform all data related to location in the environment to the local coordinates of the player. A type implemented in ORION is used to represent a position and orientation in space. This type is implemented through a matrix of homogeneous coordinates and corresponds, from a mathematical point of view, to a 3D orthonormal basis. Spatial transformations become possible in ORION when data are of this type. Such transformations are mainly changes of coordinates. We use this feature of ORION to express all the data related to a location in the coordinates system of the player.

Finally, we wish to use algorithms that accept only discrete data as inputs. Therefore we perform a discretization of data using the K-Means algorithm[3] [16].

3.3 Prediction

Earlier in this article we presented a data collection of low level behavior samples through scenarios previously performed by human players. The problem now is to replicate these behaviors individually. Thanks to ORION, we integrated easily various algorithms: KNN for navigation, FFNN for aiming and IOHMM for close combat.

Movement. For the movements of the Bot in the environment we use the CHAMELEON model proposed by Tencé et al. [24] that suggests to learn the environment with a vector quantization method called Growing Neural Gas (GNG) [13]. The learning environment algorithm rebuilds the navigation graph automatically. This information allows to know which places are reachable in the

---

[3] Data discretization is a pre-processing method that reduces the number of values for a given continuous variable by dividing its range into a finite set of disjoint intervals, and then relates these intervals with meaningful labels.

environment and to compute a path. The model has been modified to be able to learn continuously on a player without growing indefinitely but being able to extend itself if the teacher begins to use a new part of the environment. This model is called Stable Growing Neural Gas (SGNG) [25].

However, a navigation graph is not sufficient to reproduce the movements of a real player in this environment. A human player with a minimum of experience in the game has, for example, a natural tendency to strafe (move sideways) when reaching the corner of a hallway which maximizes his field of vision and allows him to quickly see if an enemy is present in the corridor. To overcome this issue we propose to submit to the SGNG, besides the player's position, his speed and direction. However, this creates an additional difficulty: one of the steps of the algorithm is to submit a data to the network and to find the two closest data. When the data represents a position in space, using the Euclidean distance to find the closest points is totally justified. But when data additionally contains speed and direction, choosing a distance to obtain the desired result becomes more difficult. The orientation is indicated in our data by a unit vector, therefore the use of a simple Euclidean distance would tend to just ignore it from the positions. We propose to perform a re-scaling by feeding the SGNG with the following distance:

$$\text{Distance}((\boldsymbol{p1}, \boldsymbol{o1}, \boldsymbol{v1}), (\boldsymbol{p2}, \boldsymbol{o2}, \boldsymbol{v2})) = d(\boldsymbol{p1}, \boldsymbol{p2}) + \alpha d(\boldsymbol{o1}, \boldsymbol{o2}) + \beta d(\boldsymbol{v1}, \boldsymbol{v2}) \quad (1)$$

where d is the regular Euclidean distance and α and β are scaling factors. This re-scaling is very simple to perform thanks to the data transformation features of ORION.

In order to choose the scale parameter related to the speed, we start from the observation that a player rarely moves backwards when simply navigating (but does, when fighting). The direction of the velocity vector is strongly correlated with the orientation. The norm of the velocity vector is also relatively constant when the player is moving (the character control being done by keyboard, which only contains binary switches). So there seems to be no need to take into account the speed when calculating the distance and we therefore choose to set β to zero. On the contrary, the orientation is essential. We expect the vector quantization performed by the SGNG to be able to contain two data having the same position but opposite orientations for example. We choose the parameter α so that criterion is met.

Figure 8 illustrates the result. On the left, the SGNG has learned the navigation graph (CHAMELEON version). On the right, the velocity vectors are shown in red and the direction vectors are shown in blue ($\alpha = 250$). We distinguish the difference between the orientation and velocity vectors in the turns, illustrating the strafing performed by the player.

To reproduce a more realistic navigation behavior, we use the vector quantization as explained above with the algorithm K-Nearest Neighbors (KNN). Figure 9 illustrates the principle. In this figure, a part of the network is displayed in (a). The point and the green arrow correspond respectively to the current position of the agent and the direction it aims in. As for the construction of the network

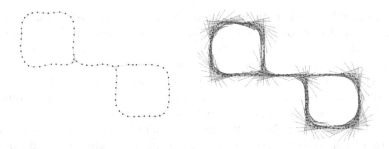

Fig. 8. SGNG Results on UT3 (Left: CHAMELEON, Right: ORION) (Color figure online)

with the SGNG, the use of KNN requires the use of a distance. Once again we need to perform a data scaling and decide to use the same distance 1. However, the parameters α and β must be selected differently: knowing the position and the desired direction of the agent, we try to get the speed and direction that the agent should adopt. So this time, we must use the velocity vector (indicating the desired direction) and not the orientation, to compute this distance. Therefore, we choose $\alpha = 0$. In order to set the parameter β, we apply the following requirement: two data should be far away if their velocities are in opposite directions, even if their position is really close.

In Fig. 9, the image (b) illustrates a possible choice of three nearest data ($k = 3$). The image (c) shows the regression then performed by weighting the closest data by the inverse of the distance to the data submitted to KNN.

Long Range Aiming. Aiming is one of the most important elements of FPS games. There are many techniques dedicated for that, and experienced players use advanced ones. The performance of the players in this activity largely determines their overall performance in the game. We chose to make it easier to reproduce this behavior, by learning it from scripted game play in which the player does not move and faces a single enemy. But an experienced player uses, at least partially, the movement for aiming. It is a very commonly seen technique

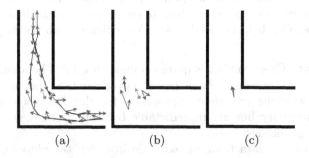

| (a) | (b) | (c) |

Fig. 9. Movement regression principle in UT3 (Color figure online)

to strafe from side to side without moving the mouse to refine the target. So we will not be able to reproduce the behavior of an experienced player. However, beginners and average players mainly perform a static aiming. So, we focus on reproducing the aiming behavior of such players.

In order to better choose a method to reproduce this behavior, we conduct an exploratory analysis. The inputs and outputs involved when performing this behavior are rather obvious. The outputs are necessarily rotational speeds (in yaw and pitch) and firing. As inputs, the position of the enemy is essential and it is likely that its speed is necessary too. By performing an exploratory analysis, we found no clear correlation between the enemy orientation and outputs. So we do not use this information as input to learn this behavior. But we found the current weapon is very important as an input. Some weapons are used by continuously pressing the fire button and others by discrete firing. In addition, the velocity of projectiles is different from weapon to weapon. Therefore, the influence of enemy speed is not the same from one weapon to another. We identify, regardless of the weapon used, the following correlations between:

- the Y position of the enemy in the player's coordinate system (right or left) and the yaw rotation speed
- the Z position of the enemy in the player's coordinate system (top or bottom) and the pitch rotation speed
- the Y enemy speed in the player's coordinate system (right or left) and the yaw rotation speed

The differences in behavior according to the weapon suggest to process different training sessions for each of them. We learn these behaviors with regression algorithms. We conducted this learning through Feed Forward Neural Network (FFNN), KNN and Support Vector Machine (SVM) algorithms. The quantitative results appeared to be quite similar from one algorithm to another but the resulting behaviors were not really correlated to the Mean Squared Error (MSE).

To compare the artificial behaviors to those of the human player, we reproduce them in our test case scenario. When collecting the data for the weapon called LinkGun, the human player scored 50 points and was killed 10 times (so the bot scores 10 points), winning the game. We have reproduced this scenario using our behavior instead of the player's. The best behavior, obtained with a multilayer perceptron, killed the bot 32 times and was eliminated 50 times, losing the game. Our behavior is therefore less efficient than the player.

Close Combat. Close combat is quite common in UT3. The arenas are often made of tortuous corridors, where one often encounters enemies at turning points. Some weapons are more appropriate for this kind of confrontation. Reflexes are necessary but an unpredictable behavior is the key to make the task of the enemy harder.

By analyzing the data from our scenario in which the player is in contact with an enemy in a small room, it was difficult to find relevant information in the context of the reproduction of this behavior. Changes in direction and jumps are common, without being obviously correlated with the slightest stimuli. The player, however, constantly tries to face the enemy. Strafing and backward runs are often used rather than the forward movement.

We tried to reproduce this behavior using standard regression algorithms that we have used for the aiming behavior. However, we have not been able to obtain convincing results with this method. This is not surprising because the exploratory analysis shows that there was no obvious dependencies between inputs and outputs. These algorithms are used to approximate a mathematical function that, for a given input always provides the same output. However, in our data set, this is clearly not the case.

Therefore, we used an Input-Output Hidden Markov Model (IOHMM) [3] to reproduce this behavior. A probabilistic model is indeed more suited to this type of data because it provides a certain unpredictability. We used a network very similar to the CHAMELEON model, applying the recommendations on variable dependencies.

3.4 Behavioral Model

We have shown how we reproduce three low level actions: shoot the enemy from afar, fighting hand-to-hand and navigating in the environment. Now we have to offer a complete behavior, using these actions in the behavioral model of ORION. The overall behavior is implemented with a Behavior Tree (BT) that we extended in order to be able to use data mining techniques. Consult ([removed for blind review]) for a complete description of the model.

Figure 10 shows a simplified version of the BT implementing the overall behavior. In this figure, the green boxes represent reference decorator nodes. They are used to name a sub-tree in order to reference that sub-tree elsewhere in the BT. They are only present here for the sake of clarity. The behavior starts by transforming the input data (discretization and change of basis). The implementation details of this node are not shown in the figure. Then, the selected

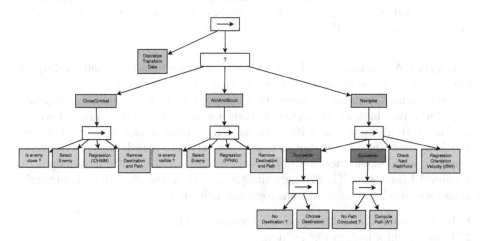

Fig. 10. ORION BT of our UT3 agent (Color figure online)

node is in charge of the choice of the low level behavior to execute. The implementation of the nodes CloseCombat and AimAndShoot is quite simple. First we test if an enemy is present (and close enough in the case of CloseCombat), following this we select the nearest enemy (by adding a variable in the execution context), we then perform the regression with the algorithm that has been trained beforehand, and finally we delete the variables that could be used by the Navigation behavior from the execution context. If no enemy is present, the Navigation behavior is executed. This behavior consist in choosing a destination (if none is already selected) and adding it in the execution context. Then, if needed, it calculates the path to get to this destination and selects the appropriate navigation point. Finally it performs regression movement with KNN. The BT is searched depth-first on each tick.

3.5   Results

Thanks to ORION tools, we tracked a player for some minutes (for each scenario as explained above) in order to learn the low level behaviors.

Subjective Analysis. Our observations from the resulting behavior are the following:

- Movement: we noticed a significant improvement with the possibility to obtain movements that reproduce the *strafing* behavior of human players at the turning points of corridors.
- Long range aiming: we noticed that the *aiming is different* for each weapon, in order to have a satisfying result we need to perform a training session for each of them.
- Close combat: The result was not entirely satisfying. The behavior of human player is *unstructured and unpredictable* which complicate its reproduction.

The resulting behaviors are promising since by watching the bot, it looks like a bot controlled by a human. In order to validate our proposition we performed a formal evaluation described in the next section.

Objective Analysis. In order to evaluate in more detail the believability of the learned movements of the bot, we carried out a study. The study consisted of two rounds of gameplay, followed by a survey. For the first round, participants played a three minutes training match against a native bot of the game to become familiar with the game and its controls. Then, players played a five minutes match against the trained bot. Finally, players were asked to answer several Likert-style questions about the the believability of this bot's movements. A four-level Likert scale (1: Strongly disagree, 2: Disagree, 3: Agree, 4: Strongly agree) was used and the questions were the followings:

1. Its movements resemble those of a human player.
2. It rotates in a human-like fashion.

3. It looks around in a human-like fashion.
4. It avoids walls in a human-like fashion.
5. The path it takes seems to be that of a human player.

Seventeen people volunteered to participate in our study. Participants do not have background knowledge about UT3 or UT-like games. Among them, 59% agreed (picked the level 3 or 4 on the Likert scale) with the question (1), 53% with the questions 2), (3) and (5), and 47% with the question (4). These results are encouraging since for all elements of movements, to the exception of wall avoidance, more than half of the participants found them believable. This allows us to point out the elements to be considered with greater importance when learning.

4   Conclusion and Future Works

We presented in this paper the ORION model. ORION permits to associate semantics to the data. This semantics is essential to enable a better understanding of the results of data mining algorithms we want to use. ORION also provides a powerful data visualization solution. Traditional charts for univariate or multivariate analysis of the data set (histogram, radar and parallel plot, etc.) are implemented and can be extended. A generic representation model allows the data to be displayed in 2 or 3 dimensions. Data visualization primitives such as points, vectors, images, texts or graphs are proposed and the model allows extensions. This allows us to display, in a generic tool, most of the data available on the UCI database.

The ORION model formalizes various techniques coming from research in data mining and proposes to gather them together according to their use. Our tool, therefore, proposes to perform tasks as diverse as data clustering, vector quantization, extraction of characteristics or prediction. The implementation of our model provides a generic approach for data analysis and data mining in the manner of WEKA or ELKI. The main contributions of our tool in relation to the latter are the management of semantics and the greater possibility of visualization. The number of available algorithms is smaller than the ones provided with WEKA but may easily be extended.

Finally, ORION also offers a behavioral model which extend the BT model. We add to this model the ability to use data mining techniques to implement complex behaviors. This model allows both online and offline learning. We illustrated the use of our model to produce AI in UT3 game using machine learning techniques. We have shown how the exploratory data analysis can provide help to make the choice of learning techniques to use. We also showed how some of the supervised and unsupervised learning algorithms can be used as part of the creation of an AI in video games. Unsupervised learning, for example, helped us to automatically reconstruct the environment and the information on how the players move in it. Supervised learning has allowed us to reproduce relatively simple actions that can be used as primitives to implement more elaborate behaviors. While the behaviors conducted with our model are neither more efficient nor more believable than behavior implemented with far more widely used methods in this industry, their implementation shows that learning methods from the world of research can help the design of AI in video games, easing the designer's workload and providing it with relevant information on human player behavior.

Many issues still need to be resolved before our behavioral model can automatically and completely learn both credible and effective behaviors without human intervention. The first opportunity for improvement is probably automatic identification of low-level behavior in the traces of the player. Temporal clustering is a difficult task but work done as part of the video segmentation, such as that of [17], seems to provide good results. While we have not been able to effectively adapt these techniques to our situations, it nevertheless seems to us that some results are encouraging and that it is worth pursuing. The addition of a reinforcement learning mechanism to our model seems to be a possible avenue for improvement. Taking into account the performance of the behavior seems a very useful source of information in order to obtain more convincing behavior. Finally, an automatic construction of BT could be seen as one of the last steps required for creating a behavior by using traces and without human intervention.

Appendix 1

See Fig. 11.

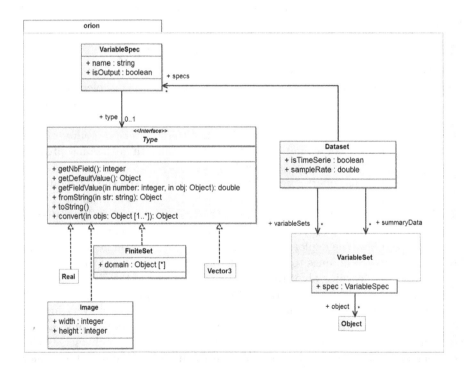

Fig. 11. ORION data model

Appendix 2

See Fig. 12.

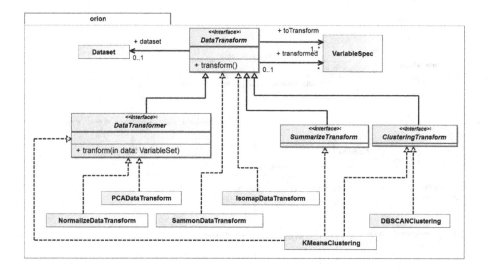

Fig. 12. ORION data transformation model

Appendix 3

See Fig. 13.

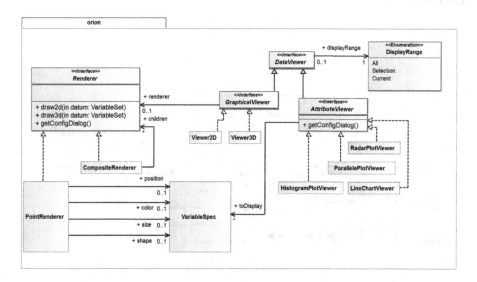

Fig. 13. ORION data visualization model

Appendix 4

See Fig. 14.

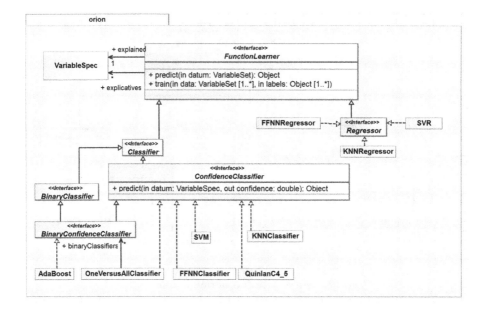

Fig. 14. ORION prediction model

Appendix 5

See Fig. 15.

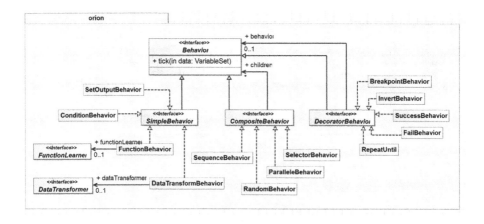

Fig. 15. ORION data mining behaviors

Appendix 6

See Fig. 16.

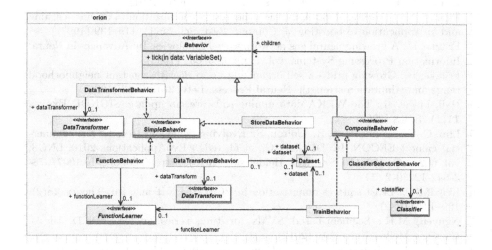

Fig. 16. ORION online learning model

References

1. Achtert, E., Kriegel, H.-P., Zimek, A.: ELKI: a software system for evaluation of subspace clustering algorithms. In: Ludäscher, B., Mamoulis, N. (eds.) SSDBM 2008. LNCS, vol. 5069, pp. 580–585. Springer, Heidelberg (2008). https://doi.org/10.1007/978-3-540-69497-7_41
2. Ankerst, M., Breunig, M.M., Kriegel, H.P., Sander, J.: OPTICS: ordering points to identify the clustering structure. ACM SIGMOD Rec. **28**, 49–60 (1999)
3. Bengio, Y., Frasconi, P.: An input output HMM architecture. In: Advances in Neural Information Processing Systems, pp. 427–434 (1995)
4. Bengio, Y., Frasconi, P.: Input-output HMMs for sequence processing. IEEE Trans. Neural Netw. **7**(5), 1231–1249 (1996)
5. Bentivegna, D.C., Atkeson, C.G., Cheng, G.: Learning from observation and practice using primitives. In: AAAI 2004 Fall Symposium on Real-life Reinforcement Learning (2004)
6. Buche, C.: Adaptive behaviors for virtual entities in participatory virtual environments. Université de Bretagne Occidentale - Brest, Habilitation à diriger des recherches (2012)
7. Buche, C., Even, C., Soler, J.: Autonomous virtual player in a video game imitating human players: the ORION framework. In: International Conference on Cyberworlds, pp. 108–113. IEEE (2018)
8. Demšar, J., et al.: Orange: data mining toolbox in python. J. Mach. Learn. Res. **14**(1), 2349–2353 (2013)

9. Ester, M., Kriegel, H.P., Sander, J., Xu, X.: A density-based algorithm for discovering clusters in large spatial databases with noise. In: Proceedings of the 2nd International Conference on Knowledge Discovery and Data mining, pp. 226–231 (1996)

10. Evans, R.: The Use of AI Techniques in Black & White (2001)

11. Freund, Y., Schapire, R.E.: A decision-theoretic generalization of on-line learning and an application to boosting. J. Comput. Syst. Sci. **55**(1), 119–139 (1997)

12. Fritzke, B.: A growing neural gas network learns topologies. In: Advances in Neural Information Processing Systems, vol. 7, pp. 625–632 (1995)

13. Fritzke, B.: Growing grid - a self-organizing network with constant neighborhood range and adaptation strength. Neural Process. Lett. **2**(5), 9–13 (1995)

14. Hall, M., et al.: The WEKA data mining software: an update. SIGKDD Explor. **11**(1), 10–18 (2009)

15. Lim, C.-U., Baumgarten, R., Colton, S.: Evolving behaviour trees for the commercial game DEFCON. In: DI Chio, C., et al. (eds.) EvoApplications 2010. LNCS, vol. 6024, pp. 100–110. Springer, Heidelberg (2010). https://doi.org/10.1007/978-3-642-12239-2_11

16. Lloyd, S.P.: Least squares quantization in PCM. IEEE Trans. Inf. Theory **28**(2), 129–137 (1982)

17. Nguyen, M.H.: Segment-based SVMs for time series analysis. Ph.D. thesis, Carnegie Mellon University (2012)

18. Quinlan, J.R.: Improved use of continuous attributes in C4.5. J. Artif. Intell. Res. **4**(1), 77–90 (1996)

19. Russel, S., Norvig, P.: Artificial Intelligence: A Modern Approach, 3rd edn. Prentice Hall, Upper Saddle River (2009)

20. Sammon, J.W.: A nonlinear mapping for data structure analysis. IEEE Trans. Comput. **C–18**(5), 401–409 (1969)

21. Schölkopf, B., Smola, A., Müller, K.R.: Nonlinear component analysis as a kernel eigenvalue problem. Neural Comput. **10**(5), 1299–1319 (1998)

22. Swayne, D.F., Buja, A.: Exploratory visual analysis of graphs in GGOBI. In: Antoch, J. (ed.) COMPSTAT 2004 — Proceedings in Computational Statistics, pp. 477–488. Physica-Verlag HD, Heidelberg (2004). https://doi.org/10.1007/978-3-7908-2656-2_39

23. Swayne, D.F., Buja, A., Lang, D.T.: Exploratory Visual Analysis of Graphs in GGobi. In: COMPSTAT, no. Dsc, pp. 477–488 (2004)

24. Tencé, F., Gaubert, L., De Loor, P., Buche, C.: CHAMELEON: a learning virtual bot for believable behaviors in video game. In: International Conference on Intelligent Games and Simulation (GAMEON 2012), pp. 64–70 (2012)

25. Tencé, F., Gaubert, L., Soler, J., De Loor, P., Buche, C.: Stable growing neural gas: a topology learning algorithm based on player tracking in video games. Appl. Soft Comput. **13**(10), 4174–4184 (2013)

26. Tenenbaum, J.B., de Silva, V., Langford, J.C.: A global geometric framework for nonlinear dimensionality reduction. Science (New York, N.Y.) **290**(5500), 2319–2323 (2000)

27. Thurau, C., Sagerer, G., Bauckhage, C.: Imitation learning at all levels of game-AI. In: Proceedings of the International Conference on Computer Games, Artificial Intelligence, Design and Education, pp. 402–408 (2004)

28. Torgerson, W.S.: Multidimensional scaling: I. Theory and method. Psychometrika **17**(4), 401–419 (1952)

29. Vapnik, V., Golowich, S.E., Smola, A.: Support vector method for function approximation, regression estimation, and signal processing. In: Advances in Neural Information Processing Systems, vol. 9, pp. 281–287 (1996)

30. Vondrick, C., Khosla, A., Malisiewicz, T., Torralba, A.: HOGgles: visualizing object detection features. In: 2013 IEEE International Conference on Computer Vision, pp. 1–8, December 2013

31. Welch, L.R.: Hidden Markov Models and the Baum-Welch Algorithm. IEEE Inf. Theory Soc. Newslett. **53**(4), 10–13 (2003)

32. Yamamoto, K., Mizuno, S., Chu, C., Thawonmas, R.: Deduction of Fighting-Game Countermeasures Using the k-Nearest Neighbor Algorithm and a Game Simulator, 4, April 2014. Ice.Ci.Ritsumei.Ac.Jp

Environment Estimation for Glossy Reflections in Mixed Reality Applications Using a Neural Network

Tobias Schwandt[(✉)], Christian Kunert, and Wolfgang Broll

Ilmenau University of Technology, Ehrenbergstraße 29, Ilmenau, Germany
{tobias.schwandt,christian.kunert,wolfgang.broll}@tu-ilmenau.de

Abstract. Environment textures are used for the illumination of virtual objects within a virtual scene. Using these textures is crucial for high-quality lighting and reflection. In the case of an augmented reality context, the lighting is very important to seamlessly embed a virtual object within the real world scene. To ensure this, the lighting of the environment has to be captured according to the current light information. In this paper, we present a novel approach by stitching the current camera information onto a cube map. This cube map is enhanced in every single frame and is fed into a neural network to estimate missing parts. Finally, the output of the neural network and the currently stitched information is fused to make even mirror-like reflections possible on mobile devices. We provide an image stream stitching approach combined with a neural network to create plausible and high-quality environment textures that may be used for image-based lighting within mixed reality environments.

Keywords: Augmented reality · Enhanced reality · Visualization · Image fusion · Computer vision · Neural networks

1 Introduction

The quality of Augmented Reality (AR) and Mixed Reality (MR) applications is a crucial factor when it comes to overall acceptance for everyday usage. Besides a sophisticated tracking and by that a high-quality geometric registration of the virtual objects, the quality and the plausibility of the visual appearance becomes more and more important.

Visual effects like indirect lighting, scattering, or refraction have been widely neglected in most AR environments although their importance is recognized to be very high. A proper illumination of objects is mainly influenced by the global lighting based on the real environment. Using the current image stream to estimate the environment lighting may be done in several ways. One of the best-known approaches is the usage of a mirroring ball, capturing the front and back scene of the environment [4]. More recent approaches make use of an RGB video stream [23, 24]. Although these approaches show some good visual results,

© Springer-Verlag GmbH Germany, part of Springer Nature 2020
M. L. Gavrilova et al. (Eds.): Trans. on Comput. Sci. XXXVI, LNCS 12060, pp. 26–42, 2020.
https://doi.org/10.1007/978-3-662-61364-1_2

they require rather complex calculations per frame to estimate the environment lighting, while still having certain limitations. The approximated background is physically incorrect, which becomes obvious, particularly with regard to flat glossy surfaces.

In this paper we present our approach, approximating the environment lighting by the following contributions:

- We provide a real-time solution for an environment lighting based on physically plausible data. Therefore, we stitch the image stream onto a cube map according to the current camera pose on runtime. The final result will be enhanced and even images obtained in bad light conditions can be stitched.
- With the aid of a neural network, we estimate completed environment textures using the partially existing information of the stitching approach. Therefore, the entire 360-degree information of the current light situation is available and a light probe may be used for diffuse and specular lighting.
- The currently stitched light information is frequently used for estimating a 360-degree illumination applying a neural network. Both are combined by blending the stitched information onto the information of the neural network. Hereby, a better and plausible reflection, even on mirror-like surfaces, can be obtained.

At the beginning we provide a review of the current state of the art in Sect. 2. Section 3 shows the usage of the image stream to stitch the information onto a cube map. In Sect. 4 we show the structure of the neural network in combination with some training results. Based on the environment stitching and the estimation of the neural network, we take a closer look at the results in Sect. 5 with a description of the combination of both techniques. In Sect. 6, we provide a short discussion including some limitations. Finally, we conclude and have a look into possible future work in Sect. 7.

2   Related Work

In this section, we will review previous work on simulating glossy reflections as part of illumination models with a focus on AR-applicable approaches. Further, we will review the stitching of multiple images to generate a panoramic image.

2.1   Illumination Reconstruction

Proper glossy reflections may be generated in several ways e.g. by synthesizing objects in the real world. For example a mirroring ball (glass or chrome ball) can be placed inside the scene to capture the background information. State et al. and Paul Debevec presented such an approach and showed the importance of a proper illumination in an AR setup [4,28]. A light probe is generated by a combination of the foreground and background information. Capturing is done using High Dynamic Range (HDR) to achieve a better quality. The approach requires a preparation of the scene and the existence and tracking of the mirroring ball.

The usage of special camera techniques is a common method to reconstruct the light. Based on a fish-eye camera, Franke captures light conditions [5]. By that, real-time glossy inter-reflections are possible between real and virtual objects. Multiple cameras with fish-eye lenses may be used to capture a more sophisticated surrounding illumination. Another approach by Rohmer et al. makes use of a stationary PC with a tracking system for computing environment lighting [22]. Hence, they can achieve highly realistic virtual reflections in real-time on mobile devices. Besides multiple cameras with fish-eye lenses and a stationary PC, a tracking system is needed to track the camera position and rotation. Gruber, Ventura and Schmalstieg showed an approach for real-time global illumination by using depth data of an RGB-D camera [8]. Based on a 3D reconstruction of the scene, enhanced with spherical harmonics, a general diffuse lighting of the scene is estimated. Using multiple cameras, fish-eye lenses [5], depth sensors [8,21], or even a stationary tracking system [22] is not applicable to most AR scenes, significantly limiting the applicability of the approaches.

Estimation of the global lighting may also be done by using a single image stream. Kán et al. show the generation of 360-degree environment map by scanning the environment before visualizing virtual objects [11]. After scanning the environment the result is filtered for different types of materials. While this approach may produce high-quality results, it is not suitable for altering light conditions and moving objects because of the static environment map. Using an image stream only Ropinski et al. presented an approach for creating a light probe every frame [23]. Each virtual object inside the scene is handled individually to create a cube map. Yao et al. enhance the generation of the environment cube map by using a sequence of input images [31]. Their approach does not allow for real-time performance nor for mutual reflections between real and virtual objects. Within some of our previous approaches [24–26], the light probe is generated each frame by using a single camera image only. While this approach achieves a pretty good approximation of the global lighting, it cannot be used when realistic reflections of the background are expected. Sometimes virtual objects are recognized as transparent objects, some parts of the environment texture are missing, or the final result shows hard edges.

Another approach is the application of scene analysis to find visually similar environment maps from a digital library. Karsch et al. presented such an approach by analyzing the scene and searching for a comparable environment map to simulate the global light [12]. This approach may be applied to indoor as well as to outdoor scenarios. However, glossy reflections are not supported due to the fact that the environment maps do not provide dynamic information. Especially objects with flat mirror-like surfaces may show artifacts not part of the real environment.

Alternatively to the usage of a single camera image, a neural network may be used to estimate the lighting. Mandl et al. show the usage of a Convolutional Neural Network (CNN) for a highly accurate estimation providing photo-realistic rendering results [16]. A CNN is trained with a set of pre-illuminated models with different spherical harmonics variations stored in a database. While rendering, depending on the current camera pose, a similar lighting from the database is used. Currently, this approach supports diffuse lighting only with a single pre-defined model that has to be trained for the database. In the case of visualizing specular reflections, Georgoulis et al. use a neural network to predict the environment by capturing reflectance maps from one to multiple real-world objects inside the current camera view [7]. The learning based approach combines the information of the image stream with the reflectance of the object on gaussian spheres. This approach allows for an environment prediction similar to our results but needs some known objects inside the scene like [4]. Another approach by Gardner et al. has trained a neural network to classify light sources and predict light intensities from a single image [6]. The neural networks are trained with Low Dynamic Range (LDR) environment maps to identify the location of the light. After this, another network is trained with HDR environment maps to identify the intensity respectively. Hereby, a HDR indoor illumination may be estimated to visualize non-mirror-like virtual objects.

2.2 Stitching

For the creation of a 360-degree panorama by image stitching, most approaches depend on multiple images with a proper amount of overlapping image information [10,29]. Modern cameras and applications consist of automatic mechanisms to stitch multiple images into panoramas. Here we want to present the most important approaches that are relevant for our work. Szeliski provides a good overview of image stitching, comparing between standard methods based on camera-, photogrammetry- and feature-based approaches [29]. Feature-based methods provide a good solution in the case of stitching multiple images together by using Random Sample Consensus (RANSAC), Scale-Invariant Feature Transform (SIFT), Speeded Up Robust Features (SURF), Harris, and other feature-based approaches [18,20]. Based on the determined feature-set, multiple images with different conditions may be automatically stitched together [1]. One problem of feature-based methods is the blurring of images as soon as objects and/or the camera are moving. The blurring of the image is addressed by [13]. They used SIFT+RANSAC and Harris+RANSAC to stitch images together. The proposed method showed a high performance when tested with a robot movement applying only one Degree of Freedom (DOF). Applying a camera movement using more then one DOF has not been tested within this approach. Dasgupta and Banerjee presented a panoramic vision based on multiple camera sources in real time [3]. Using multiple cameras is not applicable to our approach. Based on the overlapping of two adjacent images, Chew and Lian presented a panorama stitching system using only a single camera [2]. They use SURF to detect features between two images and stitch them together based on a weighted projection. Unfortunately, a real time stitching of the adjacent images has not been tested.

3   Stitching

In contrast to other approaches, additional devices such as depth sensors or fish-eye lenses are not part of our approach to ensure a high flexibility. An RGB image stream is used only for the reconstruction of the environment lighting. Therefore, most light information may only be approximated or re-interpreted. Our previous approach calculated a plausible reconstruction of the far-field illumination by extracting information of a given RGB stream. However, in that approach, the approximation of the foreground lighting is unrealistic as it is not based on any real information about the background. Nevertheless, reflections are plausible in most cases as long as the objects have no flat surface directed towards the camera. At a certain angle of reflection, as well as particular values for smoothness and metalness of the object, they look like transparent geometry. This issue may be solved using a stitching approach like the one applied in our previous work [26].

3.1   Generation

In every frame, the information of the image stream was stitched onto the environment cube map related to camera-based techniques [15,17]. In conjunction with the camera parameters, we use the current image stream, combining it with the previous images inside a cube map. Finally, a more plausible, realistic, and even physically plausible environment illumination is created. A condition for generating a plausible cube map is that each part of the environment is captured by the camera. Depending on external factors, such as the used tracking approach or Field of View (FoV) of the camera, this may be difficult to achieve. We explored that certain parts of the ceiling are never visible because the camera is mostly directed downward.

For our reconstruction approach it is assumed that the illumination relevant for the virtual objects typically is pretty far away from the camera. In fact, this is not correct because we have always an underlying desk or plane as well as other objects rather close to the virtual objects. But as a first assumption, we assume that objects are not that close. We use the camera far plane which is calculated in each frame. Therefore, the reconstruction is based on the physically-related behavior of the camera and the tracking.

3.2   Stitching Results

Rendering to the cube map is done by a single draw call using a geometry shader stage. During the execution of the application an increasingly enriched cube map is generated applying the most recent data of the camera image stream. We use alpha blending to combine the camera images. Towards the edges of the camera image, the alpha value is linearly interpolated in such way that the new image information is almost transparent at its borders. This improves the overall result of the stitching between new and previous information, making the reflections more homogeneous especially in the case of a fast moving camera, and allows for easier combination of stitched and generated data. The current texture coordinates in unit-range are used to calculate the linear interpolation from the center to the border of the screen. We apply a simple alpha blending still allowing for real-time execution on mobile devices. The first approximation may be enhanced by non real-time approaches later on.

When starting the application, no background lighting of the environment is available. Thus, most areas of the environment map used for representing the reflections, are still empty (i.e. white) as they have not yet been visible to the camera. In our previous work, we filled these areas with information from a pre-calculated Lookup Texture (LUT) allowing for a complete lighting [26]. Unfortunately, close to the edges of the image, some edges became visible because of the non-realistic (non-physical-based) approximation. As a result, the plausibility of the environmental illumination suffered severely.

However, as soon as the entire environment has been captured once, the light probe allows a realistic reflection on smooth, mirror-like surfaces. Figure 1 shows two stitched panoramas with many visible areas. The left picture represents an office environment, while the right picture is from an outdoor scene on a cloudy day. It is clearly visible that even tiny objects are identifiable inside the image but with some missing areas.

Fig. 1. The final environment consists of the far-field illumination of the surrounding scene. In this equirectangular projection, it can be observed that even small objects inside the environment are visible. Therefore, the final reflection on a mirror-like surface can provide even detail information of the environment. The top and bottom part of the projection is essentially white. A more detailed figure showing individual images at different steps is shown in Fig. 5.

Algorithm 1 shows the steps to generate the environment cube map in each frame. The result of the cube map is finally transformed to a 2D panorama with an equirectangular projection. This image data is the input to the neural network to generate the complete environment.

Algorithm 1. Rendering of the stitched environment information

---

1: Extract points   ▷ Far clipping plane extraction
2: Upload points and UV coordinates to vertex buffer
3: Set cube map as render target
4: Set web cam image as texture
5: **for all** cube faces **do**
6:      Update view-projection matrix
7:      Render geometry with alpha blending
8: **end for**
9: Transform cube map to equirectangular projection
10: Send panorama to neural network

---

4 Neural Network

Because of the limited field of view of the camera in combination with the stitching approach, some areas of the panorama might contain no valid data. If these areas are never seen by the camera, they will remain unidentified. Our previous work shows the usage of a preprocessed LUT to generate a 360-degree environment [24] enhanced by a stitching approach [26]. As already explained, some hard edges become visible because of the non-physical and non-context-sensitive behavior of the approach.

So, in the case of creating environments a more sophisticated and context dependent solution has to be used to create highly realistic results. Therefore, we use an unsupervised visual feature learning algorithm driven by context-based pixel prediction. The related approach by Pathak et al. [19] shows a CNN trained to generate the contents of an arbitrary image region. They provide that the network is even capable to do semantic in-painting. We use this semantic in-painting for filling unidentified areas in the environment lighting.

Our neural network follows the suggested implementation of Pathak et al. [19] because it shows a sufficient visual quality for context encoding after first testings. We have used the provided Generative Adversarial Networks (GAN) architecture with a generator and discriminator. The generator down-samples the input data five times while increasing the number of filters at the same time. As an activation function, we have used a leaky version of a Rectified Linear Unit with an alpha value of 0.2. After downsizing we apply a 2D convolution with 4000 filters. To create some details onto the image, we up-sample the result five times while decreasing the number of filters back to the three input channels (RGB). In this case, we have decided on the non-leaky version of Rectified Linear Unit with an alpha value of 0.2. The discriminator uses a 2D convolution resulting in a single value that indicates the quality of the panorama. We use batch normalization for all layers except the first one. A more detailed visualization is shown in Fig. 2. All values, sizes and parameters has been determined over several training iterations and shows to be sufficient in the case of performance, stability and quality.

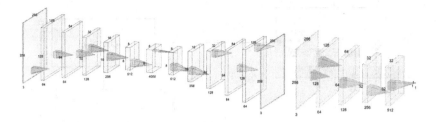

Fig. 2. On the left side, the structure of the generator is shown. The generator down-samples the input, identifies different filters and scales up to the input resolution. On the right side is the discriminator. The discriminator has the same input dimensions as the generator. In contrast to the generator, the discriminator is trained with valid and fake data to identify if the input is a real or a fake image. The images show the convolution stages and number of filters with an example resolution of 256×256 px.

For the generator, an equirectangular panorama with some undefined areas is used as input. The generator generates a new version with filled patches. Afterward, these patches are evaluated in two different ways. At first, a pixel-wise loss is calculated based on the original input and the generated one. Secondly, the discriminator decides if this panorama is real or fake. Both losses are combined to identify how good the current generator is trained. The discriminator is trained by using some generated/fake panoramas and some real one. However, the combined losses are crucial to evaluate the quality of the generator. The general pipeline is shown in Fig. 3. In addition, Fig. 4 shows further outputs of the generator regarding different light and environment situations. The visible grid structure in the second row of this figure happens because of the de-convolution (transposed convolution) steps. These checkerboard artifact is a result of overlapping in the transformation from one step to the next regarding the used kernel size and padding of our neural network architecture.

We have used the SUN360 panorama database [30] with high-resolution panorama images grouped into different place categories. Inside the database 67569 images are available with either outdoor or indoor captures. We scaled the panoramas to a resolution of 256×128 px to train our GAN. Besides, we randomly generate masks of the panorama by applying some white areas. Regarding our AR application and the stitched images we always removed the top part of the panorama. Moreover, we removed some information from the bottom part of the panorama. Lastly, we removed some random blocks in the middle of the panorama. By doing so, we comply with the behavior of a stitched image. In Fig. 4 the applied masks are visible in the first row.

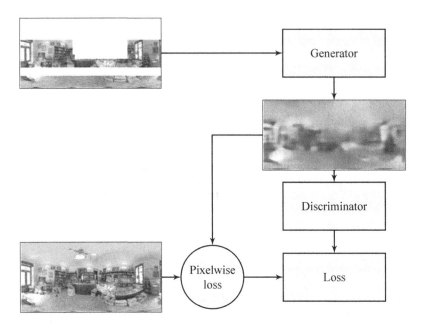

Fig. 3. The generator uses an panorama as input with some white masks that have to be filled. As an result the generator estimates an panorama that is then evaluated by the discriminator. Moreover, a pixel-wise loss is calculated depending on the real panorama and the generator output. The final loss is calculated by the output of the discriminator using a adversarial loss and the pixel-wise loss.

Fig. 4. This pullout of the results of the test data shows the panoramas with masks in the first row. Next row shows the result of the generated output by the generator. Row three is the combination between existing data and generated data. The last row shows the original panorama that has been used to estimate the error.

5   Results

The visualization using a stitched environment map, results in a good and – at the same time – realistic reflection on virtual objects. Even on smooth mirror-like surfaces, the output of the reflections is plausible.

5.1   Visualization

A closer look at the stitching approach demonstrates that a good visual quality may be achieved. Even reflections at the near-field distance are clearly visible. The network uses 256×128 RGB8 panoramas as input. White areas in the panorama are interpreted as parts that have to be filled with an in-painting. After estimating these parts with the neural network, the data is used to create the cube map. A detailed overview of the stitched input data and the output of the neural network is shown in Fig. 5. The approach works in different light situations and environments.

We combine the data of the stitching with the estimated panorama by blending to the edges of the stitched information. Hereby, the lighting is generated according to the current environment and plausible lighting is available. A transition between the stitched and generated image is not visible due to the blended environment. In Fig. 6 the combination is shown in three steps for situation "Office 1" and "Outdoor Cloudy".

Finally, a cube map (128×128 RGB16F) is used for the environment lighting of the scene. In the case of our real-time Physically Based Rendering (PBR) approach we apply an inverse tone-mapping [9] and a filtered importance sampling [14] in each frame to create a light probe (Specular: 128×128 RGB16F, Diffuse: 32×32 RGB16F). This allows diffuse and specular reflections to be displayed on virtual objects. Moreover, we use a parallax correction [27] for the reflections to reduce the issue of position dependency.

Regarding the resolution of the final output of the neural network, it can be said, that it does not have to be higher than the resolution of the light probes. In the case of increasing the resolution of the cube maps, the resolution of the output should be increased as well. Moreover, the neural network should be trained respectively. Figure 7 shows two situations of the usage of the light probe in the situation "Hallway 2" & "Outdoor Sunny" from Fig. 5. Inside this figure, a detail look on the object after tone-mapping, the combined environment, and a mapping on a perfectly glossy sphere is shown. It is clearly visible that the part of the neural network is more blurry than the part of the stitching. Regarding our Bidirectional Reflection Distribution Function (BRDF), we found that a roughness value of less than 0.8 can overcome this issue.

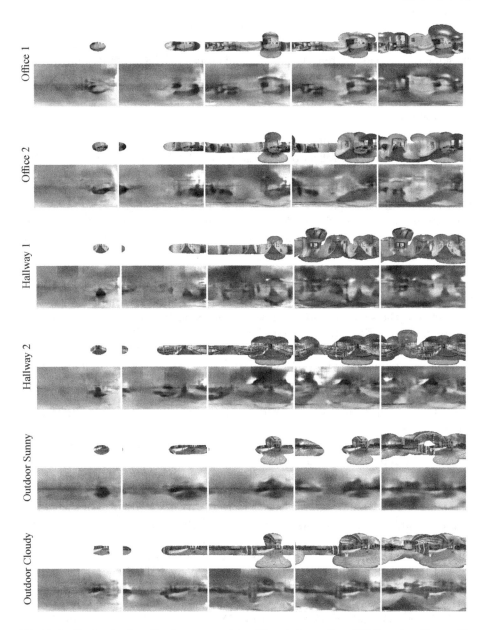

Fig. 5. In the beginning, the image consists of less information. While using the application the environment is more and more enhanced by the stitching approach. This can be seen in the first row of every situation. Based on the current information the neural network does an in-painting for white areas. This can be seen in the second row of every situation. The more information is available, the better the neural network estimates the situation.

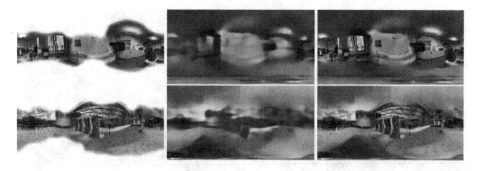

Fig. 6. The combination of the neural network and the stitching information is based on an alpha blending to the edges of the currently stitched information. This figure shows the combination for situation "Office 1" and "Outdoor Cloudy" in each row from left to right. First, the stitched information alpha value fades the information to the border. Second, the neural network calculates an environment map based on the stitched information. Finally, the information is combined to get plausible lighting information.

5.2 System

The hardware configuration used for testing was an Android Google Pixel 3 (XL) using ARCore for tracking. A self-developed in-house graphics engine has been used to render the scenarios on desktop PC and even mobile devices. The engine is based on C++ with OpenGLES 3.2 on mobile device - OpenGL 4.5 on desktop PC - and uses PBR to display virtual environments. As post effects Screen-Space Ambient Occlusion (SSAO), and Subpixel Morphological Antialiasing (SMAA) are activated.

The neural network is running on an desktop PC with an x64-based Intel Core i7-8700K @ 3.7 GHz, Nvidia GeForce RTX 2080 Ti, and 32 GB DDR4 RAM. We have used PyTorch 1.0 for Python 3.7 with CUDA 10.0 to implement, train, and test the GAN.

5.3 Performance

On the mobile device, our approach provides frame times below 32.25 ms. Filtering of the light probe - to enable physically-based reflections - is the most time-consuming step. Especially on mobile devices the filtering can be very challenging.

Sending the panorama to the server, running the neural network and sending back the final result, is almost neglectable. Especially since the generation in real-time is not a requirement because the result could be blended during execution of the application. However, measurements have shown a static computation time of 4.99 ms of the generator on the desktop PC. Running the generator on a mobile device has not been tested but should be possible in a background process.

Fig. 7. This figure shows two situations ("Hallway 2" & "Outdoor Sunny") with a screenshot, the used LDR panorama and a mapping on a glossy sphere. Inside the screenshot, a virtual head with a mirror-like material is visible. Some details behind the camera are clearly identifiable. These information has been captured some frames before. Other information that has not been seen/stitched yet is estimated by the neural network. Between the estimation and the highly detailed image, no hard edges are visible because of the smooth blending between them. The magnification shows an example area in every screenshot.

6   Discussion and Limitations

By using a stitching approach combined with a neural network it is possible to estimate a general environment lighting. The stitching allows high-quality reflections on mirror-like surfaces while the neural network estimates all the lighting that has not been seen by the camera. The estimated areas are not as sharp as the ones obtained by stitching but black areas are removed which enables a more plausible impression. Sharp edges between both approaches are prevented by blending between them. Finally, a highly realistic impression of virtual objects is possible. Even mirror-like surfaces pointing to the camera have a plausible reflection.

The reflection on the back side of the camera can only be shown with sharp details as soon as the environment was visible once. In the meantime, the reflection can only be estimated by the neural network. It has to be mentioned that the quality of the neural network is only as good as the provided information. That means as long as the stitching approach is running and a more and more sophisticated environment is generated, the better the neural network can estimate the rest of the missing information. This estimation can be insufficient because the user is not moving/rotating with the camera. In some cases, the neural network estimates a grayish environment although most of the room is dark. This happens in the case of insufficient data or training.

Due to our camera sensor, we are capturing LDR information only. This can be a problem in the case of visualizing virtual objects inside an environment with highly bright and dark areas.

7   Conclusion

Our approach shows the use of the current camera image for a dynamic enrichment of ambient lighting. For this, a stitching method was applied, which is based on the camera parameters. The geometry is converted according to the view and projection information of the camera. Finally, the current image stream is used to enhance a cube map with more information about the illumination. Missing information during runtime of the application can be avoided using an approximated 360-degree illumination by the neural network. Through this, realistic reflections of the background illumination appear on flat mirror-like surfaces. Even small details can be recognized and provide a realistic impression even with changing light conditions. The approach may be used in a wide variety of applications, such as mobile AR games, collaborative working on virtual objects in different light situations, or material editing in real-world light situation. All calculations can be carried out at run-time in a few milliseconds. So, a proper environment illumination can be achieved on runtime that could be combined with other illumination approaches.

To provide a more sophisticated estimation by the neural network, the network has to be trained with more panorama data or specialized data. We have used the full dataset of the SUN360 database which consists of more outdoor as indoor environments. Because of this, the result of the neural network tends to be better in outdoor situations. Besides, we have trained the network with nearly the same amount of pixels per mask. The network should be trained with more sophisticated masks regarding a general application lifecycle.

In general, other network settings could be tested. By changing the number of layers, using a higher resolution, considering the alpha channel to the input, or changing kernel size, artifacts like the checkerboard pattern could be overcome and the visual quality improved respectively. Moreover, to achieve a more detailed environment partial-convolution based network, de-blurring or style transfer approaches could be used.

To provide a better lighting quality, the environment should be captured or at least converted in HDR. A potential way would be a better camera or the usage of a neural network to predict an HDR environment map like Gardner et al. [6].

Another improvement would be a distinction between different light probes. By using depth data, different light probes might be filled using spatially dependent data. Thus, different lighting information depending on the surroundings can be generated and used. By a combination of the light probes, a more sophisticated visualization can be achieved.

Acknowledgment. The underlying research of these results has been partially funded by the Free State of Thuringia with the number **2015 FE 9108** and co-financed by the European Union as part of the European Regional Development Fund (ERDF).

References

1. Badra, F., Qumsieh, A., Dudek, G.: Rotation and zooming in image mosaicing. In: Proceedings Fourth IEEE Workshop on Applications of Computer Vision, WACV 1998. Institute of Electrical and Electronics Engineers (IEEE) (1998). https://doi.org/10.1109/acv.1998.732857
2. Chew, V.C.S., Lian, F.L.: Panorama stitching using overlap area weighted image plane projection and dynamic programming for visual localization. In: 2012 IEEE/ASME International Conference on Advanced Intelligent Mechatronics (AIM), pp. 250–255. Institute of Electrical and Electronics Engineers (IEEE), July 2012. https://doi.org/10.1109/AIM.2012.6265995
3. Dasgupta, S., Banerjee, A.: An augmented-reality-based real-time panoramic vision system for autonomous navigation. IEEE Trans. Syst. Man Cybern. Part A Syst. Hum. 36(1), 154–161 (2006). https://doi.org/10.1109/TSMCA.2005.859177
4. Debevec, P.: Rendering synthetic objects into real scenes. In: Proceedings of the 25th Annual Conference on Computer Graphics and Interactive Techniques - SIGGRAPH 1998, pp. 189–198. Association for Computing Machinery (ACM) (1998). https://doi.org/10.1145/280814.280864
5. Franke, T.A.: Delta voxel cone tracing. In: 2014 IEEE International Symposium on Mixed and Augmented Reality (ISMAR), pp. 39–44. Institute of Electrical & Electronics Engineers (IEEE), September 2014. https://doi.org/10.1109/ISMAR.2014.6948407
6. Gardner, M.A., et al.: Learning to predict indoor illumination from a single image. ACM Trans. Graph. 36(6), 176:1–176:14 (2017). https://doi.org/10.1145/3130800.3130891. http://doi.acm.org/10.1145/3130800.3130891
7. Georgoulis, S., Rematas, K., Ritschel, T., Fritz, M., Tuytelaars, T., Gool, L.V.: What is around the camera? In: IEEE International Conference on Computer Vision, ICCV 2017, Venice, Italy, 22–29 October 2017, pp. 5180–5188. IEEE Computer Society (2017). https://doi.org/10.1109/ICCV.2017.553
8. Gruber, L., Ventura, J., Schmalstieg, D.: Image-space illumination for augmented reality in dynamic environments. In: 2015 IEEE Virtual Reality (VR), pp. 127–134. Institute of Electrical & Electronics Engineers (IEEE), March 2015. https://doi.org/10.1109/VR.2015.7223334
9. Iorns, T., Rhee, T.: Real-time image based lighting for 360-degree panoramic video. In: Huang, F., Sugimoto, A. (eds.) PSIVT 2015. LNCS, vol. 9555, pp. 139–151. Springer, Cham (2016). https://doi.org/10.1007/978-3-319-30285-0_12
10. Kale, P., Singh, K.R.: A technical analysis of image stitching algorithm. Int. J. Comput. Sci. Inf. Technol. 6(1), 284–288 (2015)
11. Kán, P., Unterguggenberger, J., Kaufmann, H.: High-quality consistent illumination in mobile augmented reality by radiance convolution on the GPU. In: Bebis, G., et al. (eds.) ISVC 2015. LNCS, vol. 9474, pp. 574–585. Springer, Cham (2015). https://doi.org/10.1007/978-3-319-27857-5_52
12. Karsch, K., et al.: Automatic scene inference for 3D object compositing. ACM Trans. Graph. 33(3), 1–15 (2014). https://doi.org/10.1145/2602146
13. Kilbride, S., Kim, M.D., Ueda, J.: Real time image de-blurring and image stitching for muscle inspired camera orientation system. In: 2014 IEEE Workshop on Advanced Robotics and its Social Impacts (ARSO), pp. 82–87 (2014)
14. Křivánek, J., Colbert, M.: Real-time shading with filtered importance sampling. Comput. Graph. Forum 27(4), 1147–1154 (2008). https://doi.org/10.1111/j.1467-8659.2008.01252.x

15. Liao, W.-S., et al.: Real-time spherical panorama image stitching using OpenCL. In: International Conference on Computer Graphics and Virtual Reality, Las Vegas, America, July 2011
16. Mandl, D., et al.: Learning lightprobes for mixed reality illumination. In: 2017 IEEE International Symposium on Mixed and Augmented Reality (ISMAR). Institute of Electrical and Electronics Engineers (IEEE) (2017)
17. Mann, S., Picard, R.W.: Virtual bellows: constructing high quality stills from video. In: Proceedings of the IEEE International Conference on Image Processing, ICIP 1994, vol. 1, pp. 363–367. IEEE (1994)
18. Mistry, S., Patel, A.: Image stitching using Harris feature detection. Int. Res. J. Eng. Technol. (IRJET) 03(04), 2220–2226 (2016)
19. Pathak, D., Krähenbühl, P., Donahue, J., Darrell, T., Efros, A.A.: Context encoders: feature learning by inpainting. CoRR abs/1604.07379 (2016). http://arxiv.org/abs/1604.07379
20. Pravenaa, S., Menaka, R.: A methodical review on image stitching and video stitching techniques. Int. J. Appl. Eng. Res. 11(5), 3442–3448 (2016)
21. Richter-Trummer, T., Kalkofen, D., Park, J., Schmalstieg, D.: Instant mixed reality lighting from casual scanning. In: 2016 IEEE International Symposium on Mixed and Augmented Reality (ISMAR), pp. 27–36. Institute of Electrical and Electronics Engineers (IEEE), September 2016. https://doi.org/10.1109/ISMAR.2016.18
22. Rohmer, K., Buschel, W., Dachselt, R., Grosch, T.: Interactive near-field illumination for photorealistic augmented reality on mobile devices. In: 2014 IEEE International Symposium on Mixed and Augmented Reality (ISMAR), pp. 29–38. Institute of Electrical & Electronics Engineers (IEEE), September 2014. https://doi.org/10.1109/ISMAR.2014.6948406
23. Ropinski, T., Wachenfeld, S., Hinrichs, K.: Virtual reflections for augmented reality environments. In: International Conference on Artificial Reality and Telexistence, pp. 311–318 (2004)
24. Schwandt, T., Broll, W.: A single camera image based approach for glossy reflections in mixed reality applications. In: 2016 IEEE International Symposium on Mixed and Augmented Reality (ISMAR), pp. 37–43. Institute of Electrical and Electronics Engineers (IEEE), September 2016. https://doi.org/10.1109/ISMAR.2016.12
25. Schwandt, T., Broll, W.: Differential G-Buffer rendering for mediated reality applications. In: De Paolis, L.T., Bourdot, P., Mongelli, A. (eds.) AVR 2017. LNCS, vol. 10325, pp. 337–349. Springer, Cham (2017). https://doi.org/10.1007/978-3-319-60928-7_30
26. Schwandt, T., Kunert, C., Broll, W.: Glossy reflections for mixed reality environments on mobile devices. In: Cyberworlds 2018. Institute of Electrical and Electronics Engineers (IEEE) (2018). https://doi.org/10.1007/978-3-319-60928-7_30
27. Sébastien, L., Zanuttini, A.: Local image-based lighting with parallax-corrected cubemaps. In: ACM SIGGRAPH 2012 Talks, SIGGRAPH 2012, p. 36:1. ACM, New York (2012). https://doi.org/10.1145/2343045.2343094. http://doi.acm.org/10.1145/2343045.2343094
28. State, A., Hirota, G., Chen, D.T., Garrett, W.F., Livingston, M.A.: Superior augmented reality registration by integrating landmark tracking and magnetic tracking. In: Proceedings of the 23rd Annual Conference on Computer Graphics and Interactive Techniques - SIGGRAPH 1996, pp. 429–438. Association for Computing Machinery (ACM) (1996). https://doi.org/10.1145/237170.237282

29. Szeliski, R.: Image alignment and stitching: a tutorial. Found. Trend®
Comput. Graph. Vis. **2**(1), 1–104 (2006). https://doi.org/10.1561/0600000009.
http://www.nowpublishers.com/product.aspx?product=CGV&doi=0600000009
30. Xiao, J., Ehinger, K.A., Oliva, A., Torralba, A.: Recognizing scene viewpoint using
panoramic place representation. In: 2012 IEEE Conference on Computer Vision and
Pattern Recognition, pp. 2695–2702, June 2012. https://doi.org/10.1109/CVPR.
2012.6247991
31. Yao, X., Zhou, Y., Hu, X., Yang, B.: A new environment mapping method using
equirectangular panorama from unordered images. In: 2011 International Conference on Optical Instruments and Technology: Optoelectronic Measurement Technology and Systems, pp. 82010V–82010V-9. SPIE-International Society for Optical
Engineering, November 2011. https://doi.org/10.1117/12.904704

Distance Measurements of CAD Models in Boundary Representation

Ulrich Krispel[1(✉)], Dieter W. Fellner[1,2], and Torsten Ullrich[1]

[1] Fraunhofer Austria Research GmbH, Technische Universität Graz, Graz, Austria
{ulrich.krispel,torsten.ullrich}@fraunhofer.at
[2] Fraunhofer IGD, Technische Universität Darmstadt, Darmstadt, Germany
dieter.fellner@igd.fraunhofer.de

Abstract. The need to analyze and visualize distances between objects arises in many use cases. Although the problem to calculate the distance between two polygonal objects may sound simple, real-world scenarios with large models will always be challenging, but optimization techniques – such as space partitioning – can reduce the complexity of the average case significantly.

Our contribution to this problem is a publicly available benchmark to compare distance calculation algorithms. To illustrate the usage, we investigated and evaluated a grid-based distance measurement algorithm.

Keywords: Computational geometry · Computer-aided design · Benchmark · Euclidean distance

1 Motivation

In the context of Computer-Aided (Geometric) Design and Computer-Aided Manufacturing, the question of the distance between two objects arises frequently. Due to the importance of boundary representations and the fact that most objects in virtual worlds are triangulated, we focus on distance calculations between triangle meshes. In detail, this article investigates two triangle meshes

$$T^1 = \{t_1^1, \ldots, t_m^1\} \text{ and } T^2 = \{t_1^2, \ldots, t_n^2\} \tag{1}$$

with m, respectively n, triangles in three-dimensional space. If d_{ij} denotes the Euclidean distance between the triangles t_i^1 and t_j^2, the analyzed problem is to determine the minimum of all distances

The authors gratefully acknowledge the support of the Austrian Research Promotion Agency (Forschungsförderungsgesellschaft, FFG) for the research project (K-Projekt) "Advanced Engineering Design Automation (AEDA)". Furthermore, the authors would like to thank the Government of Styria for its support in the research project "Amber: Abstände, Metriken und deren Berechnung".

© Springer-Verlag GmbH Germany, part of Springer Nature 2020
M. L. Gavrilova et al. (Eds.): Trans. on Comput. Sci. XXXVI, LNCS 12060, pp. 43–63, 2020.
https://doi.org/10.1007/978-3-662-61364-1_3

$$\{d_{11}, \ldots, d_{1n}, \; d_{21}, \ldots, d_{2n}, \; d_{m1}, \ldots, d_{mn}\} \tag{2}$$

and to list all pairs (i,j) which take the minimum value.

The problem to calculate the distance between two triangulations may sound simple, but the algorithmic worst-case-complexity is quadratic: the worst case may consist e.g. of two triangle sets, which intersect in one point. In this case, the minimum distance value is zero and all pairs in Eq. (2) have to be reported. With quadratic reporting costs the overall complexity of any algorithm must have a quadratic lower bound. As real-world models have become larger in recent years, the research on fast distance calculation algorithm (in the average case) becomes more important. A preliminary version of this work has been reported in [15].

In the field of computer-aided design and computer graphics the Euclidean metric is of particular importance. In the following text the Euclidean distance function will be used, if not mentioned otherwise. Two points X, Y with corresponding position vectors $\mathbf{x}, \mathbf{y}$ have the Euclidean distance

$$d(X,Y) = ||\mathbf{x} - \mathbf{y}|| = \sqrt{(x_1 - y_1)^2 + \ldots + (x_n - y_n)^2}. \tag{3}$$

The distance between a single point X and a point set $\mathbf{Y}$ can be defined using the minimum of all distances between X and a point $Y \in \mathbf{Y}$, respectively

$$d(X,\mathbf{Y}) = \min_{Y \in \mathbf{Y}} d(X,Y). \tag{4}$$

For two point sets there are many different ways to define an oriented distance. Oriented distances are characterized by $d(\mathbf{X}, \mathbf{Y}) \neq d(\mathbf{Y}, \mathbf{X})$. Dubuisson and Jain have analyzed various distance functions [8]; among others:

$$d_H(\mathbf{X}, \mathbf{Y}) = \max_{X \in \mathbf{X}} d(X, \mathbf{Y}) \tag{5}$$

The oriented Hausdorff distance d_H, named after Felix Hausdorff, utilizes the maximum function. Taking the maximum of both oriented distances leads to a non-oriented distance; e.g. the non-oriented Hausdorff distance between $\mathbf{X}$ to $\mathbf{Y}$ takes the maximum of both oriented distances:

$$H(\mathbf{X}, \mathbf{Y}) = \max\left(\max_{X \in \mathbf{X}} d_H(X, \mathbf{Y}), \max_{Y \in \mathbf{Y}} d_H(Y, \mathbf{X})\right). \tag{6}$$

The sensible choice of the distance function heavily depends on the field of application.

A distinction can be made between symmetric and asymmetric settings, as well as between global and local criteria [25]: An asymmetric configuration, for example, is a target/actual comparison. The target surface is, by definition, error-free; each deviation is to be interpreted as an error in the actual surface. In a symmetrical configuration, both surfaces are interpreted the same way; e.g. the deviation of two height fields of a laser scans is an error that can occur in principle in both data sets. For symmetric (asymmetric) settings non-oriented (oriented) metrics are suitable. The question whether rather global or local influences should get more weight in a metric also depends on the application case.

The Hausdorff metric, for example, is suitable for the global similarity search [7]. The distance pairs calculated in this article are more suitable for installation and construction planning, where collision detection also has an important meaning; hence the listing of all pairs with minimal or even no distance.

2   Related Work

The problem described in the introduction above can be solved by several, different approaches. Each algorithm to solve the problem consists of the low-level aspects that comprehend the triangle-triangle distance function and the high-level aspects responsible e.g. for space partitioning. The high-level acceleration techniques used in this context are also used in different computer graphics application; namely in ray tracing, radiosity, collision detection, physically-based modeling & simulation, etc.

2.1   Triangle-Triangle Distance Function

The calculation of distances and intersections of two triangles in three-dimensional space is a common task in computational geometry [3]. Our implementations, which we present in Sect. 4, are based on the algorithm descriptions of "Geometric Tools for Computer Graphics" [24].

The complexity of a single evaluation of a triangle-triangle distance function depends on the spatial orientation of the two input triangles. Although the computational costs may vary significantly between different spatial configurations, we consider one function call of a triangle-triangle distance function as an elementary unit suitable to compare different space partitioning techniques.

2.2   Space Partitioning and Data Structures

Several strategies have been developed in order to reduce the number of evaluations of a triangle-triangle distance function.

Hierarchical Space Partitioning: Space partitioning approaches divide the Euclidean space in disjoint regions. These regions can be organized in hierarchical structures. Most space-partitioning systems use planes to divide space; such as BSP-tree, k^d-tree [20], r-trees [19], etc. If a space-partitioning system uses a tree, in which the leaf nodes enclose the geometric input (i.e. the set of triangles), the space partitioning is called a bounding volume hierarchy. The objects used as bounding volumes influence the desirable characteristics of a tree, such as optimal fitting of bounding volumes, tree balance, construction time or memory layout to name a few. Popular choices include spheres [22], axis-aligned bounding boxes (AABB) [1,4,31], arbitrarily-oriented bounding boxes (OBB) [9], discrete orientation polytopes (k-DOP) [14], rectangular swept spheres (RSS) [16] and slab cut balls (SCB) [17]. When building such a hierarchy, the split strategy that determines which parts of an object get placed in a sub-hierarchy are crucial for the performance of an algorithm; although there has been research on split strategies, e.g. for AABB hierarchies [14,29], no distinguished procedure has been found yet.

Non-hierarchical Space Partitioning: Space partitioning does not necessarily need to be hierarchical. A flat cell-based grid structure [11,28,30] can as well accelerate geometric operations and queries.

Object Representation: Instead of organizing the surrounding space, an acceleration technique may optimize the representation of the geometric objects to test. Level-of-Detail techniques [18], for example, reduce the number of triangles in a triangulation and therefore the costs of an evaluation of a triangle-triangle distance function. Using the necessary refinement on-demand, distance calculations and collision detection queries can be performed efficiently.

2.3   Algorithmic Analysis and Benchmarking

The problem in ray tracing, collision detection, and many other applications of computer graphics is limited comparability: the theoretical behavior of an algorithm usually depends on the (spatial) distribution of the input data, which is seldom known – especially if the input data has not been generated artificially.

Algorithmic Analysis: In the context of collision detection Weller et al. analyzed the running time of AABB trees [26]; whereas Han et al. compared cell-based and hierarchy-based contact detection algorithms [11]. Both approaches rely on conditions, which have to be met by the input data sets. In practice, these assumptions are not valid. As a consequence, different approaches to solve a computer graphics problem are often compared to each other via benchmarking.

Benchmarks: The requirements of a benchmark system for spatial index structures [10] consist of a data set of objects and the integration of different index structures, as well as the analysis and visualization of the results. While benchmarks exist for spatial database systems [23] and rigid object collision detection [27], no standardized benchmark for distance queries of CAD models exists.

3   Benchmark Description

The presented benchmark is designed to compare different distance measurement techniques regarding performance with a focus on distances between triangle soups. The benchmark consists of five test configurations. Each configuration comprehends two triangle sets with varying number of elements namely 100, 200, 500, 1 000, 2 000, 5 000, 10 000, 20 000, 50 000, and 100 000. With additional tests consisting of 200 000, 500 000, and 1 000 000 elements the benchmark is well prepared for the future.

3.1   CAD Test: Tools Data Set

This test is designed to represent an "average" CAD-application test. It consists of two CAD models with varying number of triangles in a good-case-scenario.

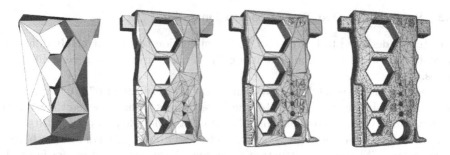

Fig. 1. The test case representing a "typical" measurement in the context of computer-aided (geometric) design consists of two tool models with varying number of faces. Figure 2 shows the corresponding second part of the distance measurement test.

The test is composed of two tools. The first one is the "Multi-Purpose Precision Maintenance Tool" by Robert Hillan – the winning entry of Future Engineers 3-D Printing in Space Tool Challenge. The data set used in this benchmark has 8 864 vertices and 17 760 faces. It has been published by NASA 3D Resources [WWW₁]. The second tool is the "Wrench" model, which has been designed on Earth and then transmitted to space for manufacture. Its 3D representation has 7 281 vertices and 14 564 faces. The original file has also been published by NASA 3D Resources [WWW₂]. Both data sets may be used for non-commercial purposes [WWW₃].

Fig. 2. The tools test measures the distance between two tools. The one shown in this Figure and the other one shown in Fig. 1. Both models have 100, 200, 500, 1000, 2000, ..., 1 000 000 faces and are rendered in wire-frame mode to outline the mesh topology. This Figure shows the resolutions 100, 500, 2000, and 10000 faces.

In order to create different test sizes both files have been reduced in size using quadric edge collapse decimation for down-sized version with 100, 200, 500, 1 000, 2 000, 5 000, or 10 000 faces. The test versions with 20 000, 50 000, 100 000, 200 000, 500 000, or 1 000 000 faces have been created by uniform mid-point subdivision (to increase the number of faces by an integral factor) and quadric edge collapse decimation (to reduce the number of faces afterwards to the predefined values) using the open source mesh processing tool MeshLab [6] (in version 1.3.4 beta 2014 [WWW₄]). The final results are illustrated in Figs. 1 and 2 showing the "Wrench" model and the "Multi-Purpose Precision Maintenance Tool" respectively.

3.2   Scale Difference Test: Rosetta Data Set

This test is designed to represent a CAD-application test of extremely different sizes. It consists of two geometric models in a configuration with very different scales.

The first model is the ESA NavCam shape model of comet 67P/Churyumov-Gerasimenko. In early August 2014, the Rosetta mission has reached the nucleus of comet 67P/Churyumov-Gerasimenko allowing a detailed mapping of its surface with the onboard imaging system, up to resolutions of 50 cm or even better in some areas. Shape reconstruction techniques have been used by Jorda et al. to build a very detailed 3D model [12]. The original 3D reconstruction has been published by the European Space Agency [WWW₅] under a Creative Commons Attribution-ShareAlike 3.0 IGO License [WWW₆]. It consists of 833 538 vertices and 1 667 073 faces. In order to have a varying model size the shape model of the comet has been reduced using quadric edge collapse decimation (see Fig. 3) to 100, 200, 500, 1 000, ..., and 1 000 000 faces. The second test model is a fixed size model; i.e. it is used in a fixed resolution in all tests. It has the shape of the space probe Rosetta built by the European Space Agency (see Fig. 4). The file has been published by NASA 3D Resources [WWW₇] for non-commercial purposes [WWW₃].

Fig. 3. The test case representing a different-scale-measurement uses the shape model of comet 67P/Churyumov-Gerasimenko in different resolutions.

Besides the model size with respect to the number of vertices and faces, the problem in this test configuration is its approximately correct scaling and positional correlation. The comet consists of two lobes connected by a narrower neck, with the larger lobe measuring ≈4.1 km and the smaller one about ≈2.6 km in diameter, whereas the space probe has a diameter of ≈32 m, respectively ≈0.032 km. Furthermore, the local circumstances are reflected by a measurement distance of ≈15 km; in short, the test scenario consists of a small, far-away part measured towards a large, big data set.

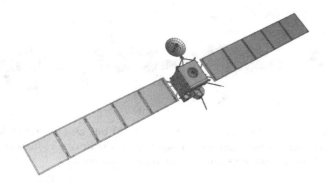

Fig. 4. The test case shows the Rosetta space probe and the comet 67P/Churyumov-Gerasimenko (shown in Fig. 3).

3.3   Tessellation Difference Test: Spirit Data Set

This test is designed to represent a heterogeneously tessellated model test. It consists of two geometric models in a configuration with elements (triangles) in very different scales.

The first model shows the Gusev Cater on Mars – the landing site of the Spirit Rover (see Fig. 5). The original data set shows 178 km × 237 km with exaggerated z-axis for better visualization effects [13]. The file with 244 447 vertices and 488 890 faces has been published by NASA 3D Resources [WWW$_8$] for non-commercial purposes [WWW$_3$]. The second test model is the corresponding Mars Exploration Rovers (Fig. 6) which has also been published by NASA 3D Resources [WWW$_9$] for non-commercial purposes [WWW$_3$].

Fig. 5. The landing site 3D data have been processed to create a heterogeneous tessellation as visible in the renderings.

In order to have a varying model size with heterogeneous tessellation the 3D models have been reduced using quadric edge collapse decimation implemented in the open source mesh processing tool MeshLab [6] (in version 1.3.4 beta 2014 [WWW$_4$]). Furthermore, randomly selected faces of the models have been subdivided using the mesh processing tool Blender [5] [WWW$_{10}$].

Besides the model size with respect to the number of vertices and faces, the problem in this test configuration is its heterogeneous tessellation; i.e. any spatial data structure to speed-up the distance calculation has to cope with non-uniform elements.

Fig. 6. Not only the landing site data set (see Fig. 5) but also the rover data set has been reduced and resampled heterogeneously. Due to the drastic reduction in size, the small size data sets <5 000 faces have severe topological errors.

3.4   Two Manifold Test: Sphere Data Set

This test comprehends a challenging configuration. Each test consists of one triangulation of random points uniformly distributed on a sphere scaled by two different factors (radius $r_1 = 1$ and $r_2 = 2$). Consequently, a test configuration with n elements in each data set should report $O(n)$ pairs with a distance ≈1; please note, in low resolution levels some "corner cutting effects" result in significantly shorter distances.

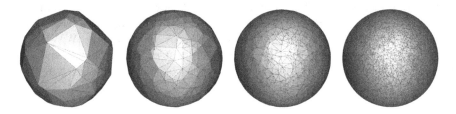

Fig. 7. The sphere test consists of triangulated points on a sphere. Each data set has been generated with 100, 200, 500, ..., input points. The resulting triangulations vary in size with approximately twice as much faces as vertices.

The test case has been generated algorithmically with two uniformly distributed random numbers $u \in [-1, 1]$ and $v \in [0, 1]$. These values are combined to three dimensional, random point $P \left(\cos v \sqrt{1 - u^2} \mid \sin v \sqrt{1 - u^2} \mid u \right)$. These random point sets have been processed with "The Quickhull algorithm for convex hulls" [2] using version 2015.2 [WWW$_{11}$]. The scaling with radius $r_1 = 1$ and $r_2 = 2$ is performed on the quickhull triangulation. Therefore, the topology of two meshes to compare is always the same. Consequently, Fig. 7 shows only one mesh per test size.

3.5   Worst Case Test: Intersection Data Set

This test comprehends a worst-case configuration; i.e. each data set consists of triangles which pass through the point $X \left(\sqrt{2} \mid -\pi \mid e \right)$. Consequently, all triangle pairs referencing a triangle in each set have distance zero. As all these pairs have to be reported, the quadratic reporting effort has the same complexity as a naive brute-force solution.

Fig. 8. The intersection test consists of triangle sets which have one intersection point in common. Each data set is rendered in solid, wireframe mode and comprehends 100, 200, 500, ..., triangles.

The test case has been generated algorithmically with a random quaternion based on a three-dimensional random vector $\mathbf{s}$. Each coordinate s_i $(i = 1, 2, 3)$ is a scalar value uniformly distributed between zero and one. With $\sigma_1 = \sqrt{1 - s_1}$, $\sigma_2 = \sqrt{s_1}$, $\theta_1 = 2\pi s_2$, and $\theta_2 = 2\pi s_3$ the random quaternion is

$$w = \sigma_2 \cos\theta_2, \quad x = \sigma_1 \sin\theta_1,$$
$$y = \sigma_1 \cos\theta_1, \quad z = \sigma_2 \sin\theta_2$$

with corresponding rotation matrix $\mathbf{R}$:

$$\begin{pmatrix} w^2 + x^2 - y^2 - z^2 & 2xy - 2wz & 2xz + 2wy \\ 2xy + 2wz & w^2 - x^2 + y^2 - z^2 & 2yz - 2wx \\ 2xz - 2wy & 2yz + 2wx & w^2 - x^2 - y^2 + z^2 \end{pmatrix}$$

This matrix is applied to three vectors

$$\mathbf{a} = \mathbf{R} \cdot \left(\cos(\tfrac{1}{6}\pi) \ \sin(\tfrac{1}{6}\pi) \ 0 \right)^{\mathrm{T}},$$
$$\mathbf{b} = \mathbf{R} \cdot \left(\cos(\tfrac{5}{6}\pi) \ \sin(\tfrac{5}{6}\pi) \ 0 \right)^{\mathrm{T}},$$
$$\mathbf{c} = \mathbf{R} \cdot \left(\cos(\tfrac{9}{6}\pi) \ \sin(\tfrac{9}{6}\pi) \ 0 \right)^{\mathrm{T}}.$$

These vectors are scaled independently by a random variable uniformly distributed between 0 and 100 (for $\mathbf{a}$), respectively between 1 and 100 (for $\mathbf{b}$ and $\mathbf{c}$). Finally, the scaled vectors $\mathbf{a}'$, $\mathbf{b}'$, and $\mathbf{c}'$ are added to the position vector of the common point X to form the position vectors of the vertices of one triangle. Figure 8 shows the final results.

3.6   Summary: Distance Measurement Benchmark

This benchmark has been designed to evaluate the performance of distance measurements in the context of computer-aided (geometric) design. It consists of five test scenarios in different extends:

- The first test scenario represents an "average" case without any special triangle configurations.
- The second scenario is a configuration with two meshes, which have a very different scale.
- The third test case is designed to evaluate a spatial data structure towards its ability to cope with extremely inhomogeneous tessellations.
- In contrast to the real-world test cases, the fourth synthetic case has a result with $O(n)$ elements. The focus of this configuration is not the spatial data structure itself, but the "hosting environment" i.e. the routines to collect and merge the results. Especially in multi-core environments, these task may turn out to be a bottleneck.
- The fifth and last scenario is also a synthetic case; furthermore, it is a pathological case which requires a minimum runtime of $O(n^2)$ for any spatial acceleration structure. This worst-case scenario may not be the most important test scenario due to its reduced practical relevance. Nevertheless, it often points out weaknesses in implementations.

All these test cases are available in different sizes – namely with 100, 200, 500, 1 000, 2 000, 5 000, 10 000, 20 000, 50 000, 100 000, 200 000, 500 000, and 1 000 000 elements. The models are stored in Wavefront .obj format [21], each file compressed using gzip. Altogether, the benchmark consists of 118 compressed files with a total size of 388 MB.

All used models are free to use for non-commercial purposes and are published online at:

https://github.com/FhA-VC

4   Distance Measurement

As stated in the introduction in Sect. 1, the input data set consists of two so-called triangle soups; i.e. two sets of triangles (cf. Eq. (1)) without any additional information. The distance measurement problem is to identify the minimum value of all triangle-triangle distances (cf. Eq. (2)) and to list all triangle pairs, which take this minimum value.

We use the presented benchmark to investigate and evaluate an algorithm that is based on a regular grid which is traversed to determine the distance.

4.1   Distance Algorithm

The grid-based acceleration structure solves the distance measurement problem in two steps: at first the set-up step builds the grid structure; afterwards the grid is used to measure the triangle distances.

Building the Acceleration Structure: For each triangle set, the acceleration structure determines the axis-aligned bounding box (AABB) and the preferred cell size. The choice of the preferred cell size is a very important factor considering memory footprint and performance. If the cell size is too large, all triangles may belong to the same cell. In this case the acceleration structure has no impact (compared to the naive approach). If the cell size is too small with more occupied cells than triangles, the impact may even be negative. The question of a "good" heuristic is dealt with in a separate section after the algorithmic description (Fig. 9).

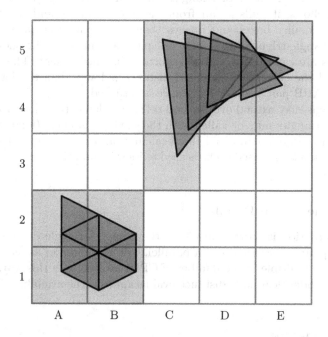

Fig. 9. The grid-based acceleration structure sorts all triangles in cells and uses the common grid structure to estimate distance bounds. In "average" distance calculation scenarios only the triangles of a few cells have to be inspected.

In a simple point-to-triangle set distance query, the preferred cell size can be used directly. If the distance between two (or more) triangle sets has to be determined, all involved triangle sets have to agree on a common cell size. Our implementation uses the median of all preferred cell sizes.

Distance Test: In order to efficiently perform the distance test, the algorithm should reduce the number of necessary triangle-triangle tests. A grid cell represents an efficient mapping of a spatial extend to a group of triangles, and thus allows to give an approximate distance by calculating the distance to the cells bounding box. The basic idea combines two different approaches: on the one hand it uses a sorted list of all possible pairs of occupied cells of the two triangle sets. Each pair has a lower bound; the distance between two triangles cannot be shorter than the distance between the cells they are hashed to.

Because this list can get quite large, it is not feasible to create and sort this list in every case. Therefore, on the other hand, we propose a hierarchical traversal, similar to AABB trees. In this case, the algorithm maintains a priority queue of sorted AABB pairs, where each AABB contains a set of occupied grid cells. The queue is sorted by the distance between the two AABB's, which is a lower bound for the distance between triangles inside the boxes. In the beginning the queue contains one AABB pair with all the cells of both triangle sets. Each pair also stores the number of triangles for each AABB. Then, the algorithm proceeds as follows; it pulls a pair from the priority queue and subdivides the AABB's into smaller boxes and pushes them back onto the queue, until the number of triangle-triangle tests in a cell pair falls under a threshold, in this case the tests are carried out and the shortest distance is kept. This process is repeated until the currently calculated distance is lower than the lower bound of the next AABB pair – and the distance is returned.

As a triangle may extend over several cells, the elementary distance function between two triangles may be called with the same parameters (triangle indices) more than once. In order to avoid the recalculation of already calculated, intermediate results, a hash-based cache is used to identify and skip already calculated triangle pairs.

4.2   Implementation Details

The implementation is written in C++, the triangle-triangle distance calculations of leaf pairs are carried out in parallel. The threshold of elements per leaf node is set to a multiple of the number of CPUs available. In this way, the work load of a leaf node↔leaf node distance evaluation uses the available CPU power efficiently.

4.3   Naive Algorithm

For comparison, a brute-force approach simply calculates all distances of all triangle pairs (i, j), for example, with two nested loops. We implemented this approach for two purposes:

1. Due to its simplicity, this implementation is robust and its correctness can be verified easily. As a consequence, this implementation has been used to verify the results of the previously described algorithm, but it is not considered to be a practical solution.
2. Furthermore, this implementation is used to optimize "low-level" routines, namely the triangle-triangle distance function, towards performance.

5   Analysis

The previously described data grid has a clear structure, is easy to implement and shows the potential to match or even undercut the runtimes of tree-based algorithms [15]. Unfortunately, the performance of grid-based algorithms depends on the choice of the optimal grid size. Unfortunately, a reasonable heuristic to determine a "good" grid size is not known – to the best of our knowledge.

Consequently, this analysis determines the optimal grid size for each test case in a brute-force manner. The goal is to derive a new heuristic.

The presented benchmark is executed on a 64-bit Linux system with 56 cores. All binaries have been compiled with GCC (version 4.9.2) with optimization level 2.

5.1   Elementary Distance Function

The naive implementation is not included in the performance analysis. Due to its simplicity and its brute-force nature, its running time does not depend on the spatial orientation of the input triangle sets. If the triangle sets comprehend n and m triangles, the naive approach always performs $n \cdot m$ calls of the triangle-triangle distance function. On the system described above the single-core implementation needs

$$0.01676\,\text{ms} \pm 0.00715\,\text{ms}$$

time to execute a single distance function call – based on all tests of the benchmark (average value $\pm$ standard deviation).

5.2   CAD Test: Tools Data Set

The tools test is the "average" application test. The objects are similar in terms of spatial extend and tessellation density. In detail, the two test objects have the axis-aligned bounding boxes

$$\begin{pmatrix} -34.4587 \\ -50.0000 \\ 0.0000 \end{pmatrix} \leftrightarrow \begin{pmatrix} 37.5064 \\ 49.9927 \\ 13.2607 \end{pmatrix}$$

and

$$\begin{pmatrix} 69.6803 \\ -46.6632 \\ -58.3412 \end{pmatrix} \leftrightarrow \begin{pmatrix} 95.9813 \\ -12.4025 \\ 55.4877 \end{pmatrix},$$

respectively. The minimum range Δ_{min} and the maximum range Δ_{min} of both objects along each axis define the sampling interval of the grid size tests. The minimum range $\Delta_{\text{min}} = 13.2607$ is at the first test object along the z-axis; the maximum range $\Delta_{\text{max}} = 113.8289$ is at the second test object along the z-axis. The sampling interval is

$$I = \left[\frac{1}{10}\Delta_{\text{min}} \, , \, \frac{1}{2}\Delta_{\text{max}} \right] \tag{7}$$

The upper boundary is chosen by the constraint that even the largest object should be divided into several cells; the lower bound is chosen on the basis of pre-experiments.

The chart in Fig. 10 plots the results of the grid-based distance calculation for the tools test with test sizes 100, 200, 500, 1 000, 2 000, and 5 000 elements. Each single plot shows the needed time of the distance queries without the needed time to build the acceleration structures (y-axis in seconds) for a chosen grid cell size (x-axis). The different test sizes are plotted in different shades of blue.

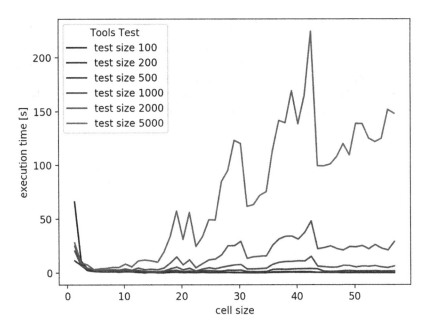

Fig. 10. The tools test of the benchmark represents the "average" application scenario. This graph plots the needed time of a distance query over the grid cell size of the acceleration structure. (Color figure online)

The plots show the wide range of results (between less than one second and more than 200 s for the same data set) and the very sensitive dependence on the cell size. Despite the strong fluctuations, the visualization shows a frequently occurring "parabolic" pattern: the optimum cell size is located between the size of the "too small cells" and the "too large cells".

5.3 Scale Difference Test: Rosetta Data Set

The Rosetta test is designed to represent an application test with two geometric models in a configuration with very different scales. The axis-aligned bounding box of the Rosetta space probe is

$$\begin{pmatrix} 12.2253 \\ 4.7607 \\ 7.2448 \end{pmatrix} \leftrightarrow \begin{pmatrix} 12.2554 \\ 4.7739 \\ 7.2535 \end{pmatrix},$$

whereas the bounding box of the comet is

$$\begin{pmatrix} -2.38839 \\ -1.77725 \\ -1.65254 \end{pmatrix} \leftrightarrow \begin{pmatrix} 2.56604 \\ 1.83242 \\ 1.52035 \end{pmatrix}.$$

The resulting sampling interval for the cell sizes is

$$I = \left[\frac{1}{10} \cdot 0.00871 \ , \ \frac{1}{2} \cdot 4.95443 \right] = [0.000871 \ , \ 2.477215].$$

The results are plotted in Fig. 11 with timings in seconds. Please note, that in contrast to the first test in which both test models have varying sizes, the Rosetta test has a fixed size model of 17 432 triangles and a model with varying size.

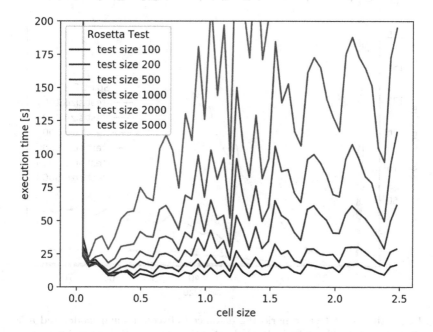

Fig. 11. The Rosetta tests with large differences in scale show extreme spikes and numerous fluctuations, most likely caused by grid discretization.

The results of the Rosetta test set (cf. Fig. 11) illustrate both the importance of a good heuristic and the difficulty in choosing the perfect cell size.

5.4   Tessellation Difference Test: Spirit Data Set

The Spirit test represents a heterogeneously tessellated model test. Based on the axis-aligned bounding boxes of the test objects

$$\begin{pmatrix} -57.6352 \\ -1.1543 \\ -83.3180 \end{pmatrix} \leftrightarrow \begin{pmatrix} 93.7512 \\ 10.0663 \\ 54.4408 \end{pmatrix} \quad \text{and} \quad \begin{pmatrix} -1.1399 \\ -0.0030 \\ -0.9749 \end{pmatrix} \leftrightarrow \begin{pmatrix} 1.1403 \\ 1.5815 \\ 0.7129 \end{pmatrix},$$

this analysis step uses the sampling interval

$$I = [0.1653, \quad 75.6982].$$

Due to the heterogeneous tessellation many triangles extend across many grid cells. Nevertheless, the results plotted in Fig. 12 do not show any conspicuities.

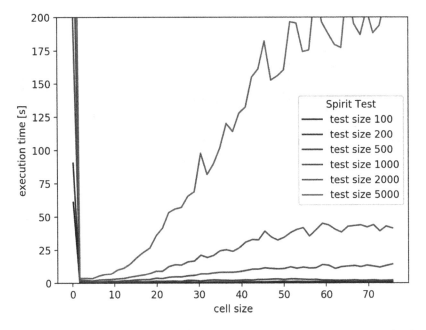

Fig. 12. In the Spirit test scenario the distance between a small model and a large, irregularly triangulated model is determined. The optimal cell size provides extremely short runtimes, whereas a poor choice leads to an extreme slowdown of the algorithm.

For the grid-based approach, inhomogeneous tessellation does not pose a problem, since a triangle registered into several cells does not lead to more triangle tests, since all tests already executed are cached and discarded when called again. This behavior is also visible in Fig. 13. It shows the number of performed elementary triangle-to-triangle distance calculations. Due to the used cache, the number of elementary tests never exceeds the "brute-force-limit" of a naive all-against-all test.

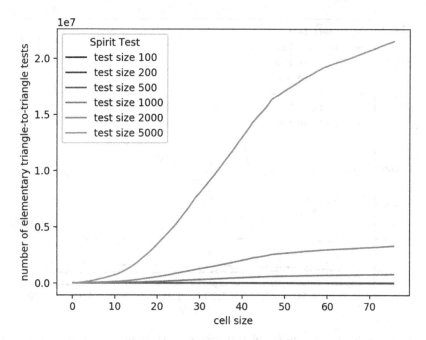

Fig. 13. The number of needed elementary triangle-to-triangle distance calculations increases with the size of the grid cells. For small cell sizes, the number of tests is minimal, but the organizational overhead then dominates the runtime (see Fig. 12).

5.5   Two Manifold Test: Sphere Data Set

In contrast to the real-world test cases, the Sphere test is a synthetic case. Its results can be seen in the charts plotted in Fig. 14. In this scenario the triangle sets have a regular distribution and many triangle pairs $(O(n))$ have to be evaluated due to a similar distance. As a consequence, this scenario is a challenge for any acceleration structure. The measured timings are now in the range of minutes.

5.6   Worst Case Test: Intersection Data Set

As the Intersection test is a pathological case with a minimum runtime of $O(n^2)$ for any spatial acceleration structure, this worst-case scenario is not suited for performance benchmarks. The ratio of calls of the triangle-triangle distance function with respect to the naive implementation is always 100%. Compared to the needed time to evaluate the distance queries, the time to build the acceleration structure is a minor factor.

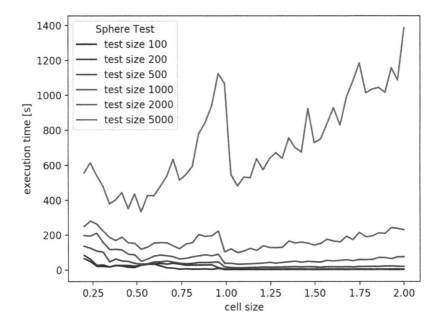

Fig. 14. The sphere test consists of two spheres with radii $r_1 = 1$ and $r_2 = 2$. In contrast to the previous real-world data sets, the timings of this synthetic scenario show less fluctuations: the needed time to calculate the distance still depends on the grid cell size, but the sensitivity is smaller by several orders of magnitude.

6   Discussion

In this paper, we presented a benchmark for systematic evaluation of distance measurement algorithms for virtual reality applications and computer-aided design constructions. The benchmark consists of five test scenarios that cover different evaluation aspects: the average case, inhomogeneous scale, inhomogeneous tessellation and two synthetic configurations that test for implementation bottlenecks and correctness.

6.1   Conclusion

In order to illustrate the usefulness of a good benchmark, we used the test sets to analyze a grid-based acceleration structure: a regular grid. The evaluation results indicate that grid-based spatial hashing is a powerful technique, but it also pinpoints to its sensitivity on the cell size. In pre-experiments and previous work [15] we already noticed the need to investigate a good heuristic to choose a reasonable grid size. In this article we determined the optimal cell sizes for all test cases with various test sizes in the interval

$$I = \left[\frac{1}{10} \Delta_{\min}, \ \frac{1}{2} \Delta_{\max} \right] \tag{8}$$

with the minimum range $\Delta_{\min}$ and the maximum range $\Delta_{\min}$ of both objects along each coordinate axis. Within this interval, the optimal parameter can be expressed by a single value $\alpha \in [0,1]$:

$$\frac{1}{10}\Delta_{\min} + \alpha \cdot \left(\frac{1}{2}\Delta_{\max} - \frac{1}{10}\Delta_{\min}\right).   \tag{9}$$

For the test included in this benchmark, the value $\alpha = 8.5\%$ seems to be a reasonable choice as the subintervals with the shortest runtimes of all tests are located at $\alpha = 8.5\% \pm 4.5\%$. This result leads to a new heuristic to choose a reasonable grid cell size. It is a valuable contribution to spatial hashing from which all users of spatial hashing and grid-based algorithms will benefit.

All in all, the benchmark demonstrated its capability of evaluating distance measurement algorithms and identifying implementation problems; it is a valuable contribution to the field of computer-aided geometric design. The benchmark is available for download and public use at:

https://github.com/FhA-VC

The research on spatial data structures will benefit from a publicly available test set to use and to compare to.

6.2   Future Work

Using this benchmark, our future work will focus on the influence of the spatial distribution of the input data sets on the runtime behaviour of acceleration structures; this research involves, amongst other aspects, the validation and testing of the new heuristics for regular grid cell sizes and its comparison to split strategies used in tree-based bounding volume hierarchies.

Furthermore, the benchmark will be extended in several ways:

– For each test scenario we will add more model sequences to increase representativeness.
– A completely new real-world test scenario will cover the use case "laser scanning", including range scanned data sets.
– Additionally, we will extend the synthetic test cases by configurations with parallel planes.

The updated benchmark will be published at the web resource mentioned above.

Web Resources

[WWW$_1$] https://nasa3d.arc.nasa.gov/detail/mpmt
[WWW$_2$] https://nasa3d.arc.nasa.gov/detail/wrench-mis
[WWW$_3$] https://www.nasa.gov/multimedia/guidelines/index.html
[WWW$_4$] http://www.meshlab.net

[WWW$_5$]
http://blogs.esa.int/rosetta/2015/08/13/a-shape-model-whats-that
[WWW$_6$] https://creativecommons.org/licenses/by-sa/3.0/igo/
[WWW$_7$] https://nasa3d.arc.nasa.gov/detail/eoss-rosetta
[WWW$_8$] https://nasa3d.arc.nasa.gov/detail/SpiritLanding
[WWW$_9$] https://nasa3d.arc.nasa.gov/detail/spirit
[WWW$_{10}$] https://www.blender.org
[WWW$_{11}$] http://www.qhull.org

References

1. Alliez, P., Tayeb, S., Wormser, C.: 3D fast intersection and distance computation. CGAL User and Reference Manual (2016)
2. Barber, C.B., Dobkin, D.P., Huhdanpaa, H.: The quickhull algorithm for convex hulls. ACM Trans. Math. Softw. **22**(4), 469–483 (1996)
3. de Berg, M., Cheong, O., van Kreveld, M., Overmars, M.: Computational Geometry: Algorithms and Applications. Springer, Heidelberg (2008). https://doi.org/10.1007/978-3-540-77974-2
4. van den Bergen, G.: Efficient collision detection of complex deformable models using AABB trees. J. Graph. Tools **2**, 1–13 (1997)
5. Blender Documentation Team: Blender 2.78 Manual. Blender Documentation Team (2017)
6. Cignoni, P., Callieri, M., Corsini, M., Dellepiane, M., Ganovelli, F., Ranzuglia, G.: MeshLab: an open-source mesh processing tool. In: Proceedings of the Eurographics Italian Chapter Conference, vol. 6, pp. 129–136 (2008)
7. Cignoni, P., Rocchini, C., Scopigno, R.: Metro: measuring error on simplified surfaces. Comput. Graph. Forum **17**(2), 167–174 (1998)
8. Dubuisson, M.P., Jain, A.K.: A modified Hausdorff distance for object matching. In: Proceedings of the 12th IAPR International Conference on Pattern Recognition, vol. 1, pp. 566–568 (1994)
9. Gottschalk, S., Lin, M.C., Manocha, D.: OBBTree: a hierarchical structure for rapid interference detection. In: Proceedings of the Annual Conference on Computer Graphics and Interactive Techniques, vol. 23, pp. 171–180 (1996)
10. Gurret, C., Manolopoulos, Y., Papadopoulos, A.N., Rigaux, P.: The BASIS system: a benchmarking approach for spatial index structures. In: Böhlen, M.H., Jensen, C.S., Scholl, M.O. (eds.) STDBM 1999. LNCS, vol. 1678, pp. 152–170. Springer, Heidelberg (1999). https://doi.org/10.1007/3-540-48344-6_9
11. Han, K., Feng, Y.T., Owen, D.R.J.: Performance comparisons of tree-based and cell-based contact detection algorithms. Int. J. Comput. Aided Eng. Softw. **24**, 165–181 (2007)
12. Jorda, L., et al.: The Shape of Comet 67P/Churyumov-Gerasimenko from Rosetta/Osiris Images. AGU Fall Meeting, vol. 47, p. P41C-3943 (2014)
13. van Kan Parker, M., Zegers, T., Kneissl, T., Ivanov, B., Foing, B., Neukum, G.: 3D structure of the Gusev Crater region. Earth Planet. Sci. Lett. **294**, 411–423 (2010)

14. Klosowski, J.T., Held, M., Mitchell, J.S.B., Sowizral, H., Zikan, K.: Efficient collision detection using bounding volume hierarchies of k-DOPs. IEEE Trans. Visual Comput. Graphics **4**, 21–36 (1998)
15. Krispel, U., Fellner, D.W., Ullrich, T.: A benchmark for distance measurements. In: Proceedings of the International Conference on Cyberworlds, pp. 120–125 (2018)
16. Larsen, E., Gottschalk, S., Lin, M.C., Manocha, D.: Fast distance queries with rectangular swept sphere volumes. In: Proceedings of the IEEE International Conference on Robotics and Automation, vol. 4, pp. 3719–3726 (2000)
17. Larsson, T., Akenine-Müller, T.: Bounding volume hierarchies of slab cut balls. Comput. Graph. Forum **28**, 2379–2395 (2009)
18. Luebke, D., Watson, B., Cohen, J.D., Reddy, M., Varshney, A.: Level of Detail for 3D Graphics. Morgan Kaufmann, Burlington (2002)
19. Manolopoulos, Y., Nanopoulos, A., Papadopoulos, A.N., Theodoridis, Y.: R-Trees: Theory and Applications. AI&KP. Springer, London (2006). https://doi.org/10.1007/978-1-84628-293-5
20. Moore, A.W.: An introductory tutorial on k^d-trees. Technical report, Computer Laboratory, University of Cambridge, vol. 209, pp. 1–20 (1991)
21. Murray, J.D., vanRyper, W.: Encyclopedia of Graphics File Formats, 2nd edn. O'Reilly Media, Sebastapol (1996)
22. Quinlan, S.: Efficient distance computation between non-convex objects. In: Proceedings of the IEEE International Conference on Robotics and Automation, vol. 4, pp. 3324–3329 (1994)
23. Ray, S., Simion, B., Brown, A.D.: Jackpine: a benchmark to evaluate spatial database performance. In: Proceedings of the IEEE International Conference on Data Engineering, vol. 27, pp. 1139–1150 (2011)
24. Schneider, P., Eberly, D.H.: Geometric Tools for Computer Graphics. Morgan Kaufmann, Burlington (2002)
25. Ullrich, T., Settgast, V., Fellner, D.W.: Abstand: distance visualization for geometric analysis. Project Paper Proceedings of the Conference on Virtual Systems and MultiMedia Dedicated to Digital Heritage (VSMM), vol. 14, pp. 334–340 (2008)
26. Weller, R., Klein, J., Zachmann, G.: A model for the expected running time of collision detection using AABB trees. In: Proceedings of the Eurographics Symposium on Virtual Environments, vol. 12, pp. 11–17 (2006)
27. Weller, R., Sagardia, M., Mainzer, D., Hulin, T., Zachmann, G., Preusche, C.: A benchmarking suite for 6-DOF real time collision response algorithms. In: Proceedings of the ACM Symposium on Virtual Reality Software and Technology, vol. 17, pp. 63–70 (2010)
28. Yang, S., Yong, J.H., Sun, J.G., Gu, H.J., Paul, J.C.: A cell-based algorithm for evaluating directional distances in GIS. Int. J. Geogr. Inf. Sci. **24**, 577–590 (2010)
29. Ytterlid, R., Shellshear, E.: BVH split strategies for fast distance queries. J. Comput. Graph. Tech. (JCGT) **4**, 1–25 (2015)
30. Zalik, B., Kolingerova, I.: A cell-based point-in-polygon algorithm suitable for large sets of points. Comput. Geosci. **27**, 1135–1145 (2001)
31. Zomorodian, A., Edelsbrunner, H.: Fast software for box intersections. In: Proceedings of the Annual Symposium on Computational Geometry, vol. 16, pp. 129–138 (2000)

An Immersive Virtual Environment for Visualization of Complex and/or Infinitely Distant Territory

Atsushi Miyazawa[⊠], Masanori Nakayama, and Issei Fujishiro

Keio University, 3-14-1 Hiyoshi, Kohoku-ku, Yokohama, Kanagawa 223-8522, Japan
{miyazawa,nakayama,fuji}@fj.ics.keio.ac.jp

Abstract. With the advent of the high-performance graphics and networking technologies that enable us to create virtual worlds networked via the Internet, various virtual environments have been developed to support mathematics education at around the beginning of the 21st century. In the environments that have been inherently two- or three-dimensional Euclidean, students have discovered and experienced mathematical concepts and processes in almost the same ways that they can do in real life. Although elementary mathematics, for instance, calculus and linear algebra, plays an essential role in areas of understanding and knowledge to solve real-world problems, there are traditionally three general areas in pure mathematics for advanced problem-solving techniques: algebra, analysis, and geometry. So, using Virtual Reality (VR) as a general and advanced tool for mathematics education to support students not only in the primary and secondary, also in higher education, the virtual environment ideally provides a wide variety of mathematical domains as possible. We present an immersive virtual environment that allows the user to set environmental limits beyond three-dimensional Euclidean space. More specifically, by setting the limits to n-dimensional complex projective space, an element of both complex and infinitely distant domain can be naturally visualized as a recognizable form in the Euclidean 3-space. The problem here is that the higher the level of mathematics, the more the visualization method tends to become abstract that only experts with advanced degrees can fathom. We also show how our figurative approach is essential for bridging the gap between elementary and more advanced mathematical visualizations.

Keywords: Virtual Reality · Mathematics education · Complex projective space · Mathematical visualization · Dimensionality reduction · Figuration

1 Introduction

One of the basic properties of Virtual Reality (VR) is having a high degree of design freedom available for virtual world creation. Since the mid-1980s when

© Springer-Verlag GmbH Germany, part of Springer Nature 2020
M. L. Gavrilova et al. (Eds.): Trans. on Comput. Sci. XXXVI, LNCS 12060, pp. 64–78, 2020.
https://doi.org/10.1007/978-3-662-61364-1_4

the prototype of the immersive virtual environment was built at NASA's Ames Research Center [1], VR has been greatly expected to be an effective way of teaching complex concepts to students. With the advent of the high-performance graphics and networking technologies that enable us to create virtual worlds networked via the Internet, various virtual environments have been developed to support mathematics education at around the beginning of the 21st century. *CyberMath* [2] is an avatar-based shared virtual environment built like a museum with a virtual lecture hall in its center. It is suitable for exploring and teaching mathematics in situations where both teachers and students coexist or are physically separated. The paper made a big contribution to identifying the three main issues (i.e., immersion, collaboration and teaching strategies, and realism) to be discussed. The *VRMath* [3] system is a non-immersive, online application that utilises desktop VR combined with the power of a Logo-like programming language to facilitate learning of 3D geometry concepts nd processes. VRMath enables children to manipulate objects and write programs to create objects in a VRML environment. *Construct3D* [4] is a three-dimensional dynamic geometry construction tool that can be used in high school and university education. It uses Augmented Reality (AR) to provide a natural setting for face-to-face collaboration of teachers and students. Development of the Augmented reality 3D geometry system [5] was inspired by the real difficulties students have when studying descriptive geometry and interpreting technical drawings. The system utilizes ARToolkit for tracking and user interaction. In the environments that have been inherently two- or three-dimensional Euclidean, students have discovered and experienced mathematical concepts and processes in almost the same ways that they can do in real life. Although elementary mathematics, for instance, calculus and linear algebra, plays an essential role in areas of understanding and knowledge to solve real-world problems, there are traditionally three general areas in pure mathematics for advanced problem-solving techniques: algebra, analysis, and geometry. So, using VR as a general and advanced tool for mathematics education to support students not only in the primary and secondary, also in higher education, the virtual environment ideally provides a wide variety of mathematical domains as possible (e.g., complex and/or infinitely distant ones). We present an immersive virtual environment that allows the user to set environmental limits beyond three-dimensional Euclidean space. More specifically, by setting the limits to n-dimensional complex projective space, an element of both complex and infinitely distant domain can be naturally visualized as a recognizable form in the Euclidean 3-space. The problem here is that the higher the level of mathematics, the more the visualization method tends to become abstract that only experts with advanced degrees can fathom. We also show how our figurative approach [6] is essential for bridging the gap between elementary and more advanced mathematical visualizations.

In advanced mathematics, any problems are defined and solved in the n-dimensional complex projective space. Expanding the range of numbers from real to complex is often called *complexification*. For instance, the complex wave is known to play an essential role in quantum mechanics—it can explain natural

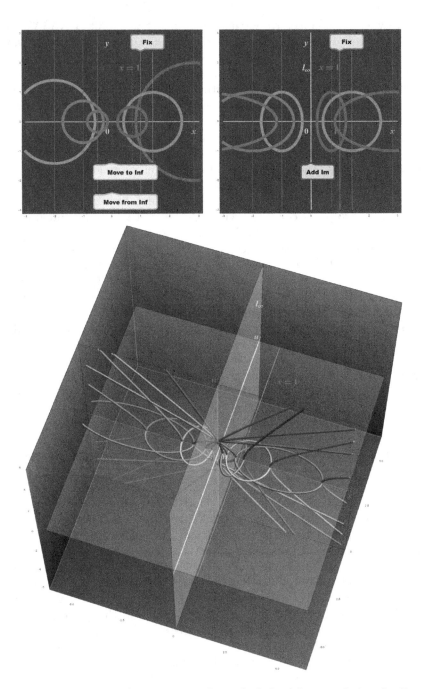

Fig. 1. Locating two circular-points-at-infinity (red dots) by translating the line-at-infinity to $x = 0$ and converting it from real to complex (from top to bottom). (Color figure online)

phenomena very well. The optical evanescent wave, also known as "imaginary light", is derived as the plane wave solution of the Maxwell equation (written in complex exponential function form). In other words, the rules of nature can be argued to have been written with logic and using complex numbers. Having various idioms for visualizing the properties of a function of at least one complex variable is important to unravel the mysteries of the real world. We are accustomed to the convenience of visualizing the overall behavior of a function by means of its graph. In the case of a complex function, however, this approach has not seemed workable since Carl Friedrich Gauss clarified the concept of the complex number plane in the 19th century, as four dimensions are required to depict a pair of complex numbers (an input $z = x + iy$ and the output of a function $f(z) = u + iv$). Charles Howard Hinton who coined the official name for an unraveled hypercube, a tesseract, in his 1904 book [7], contributed to the popularization of higher-dimensional objects using three methods: by examining their shadows, their cross sections, and their unravellings [8]. Figure 2 shows two ways: with a cross-section (left) and by looking at the shadow cast on the third dimension (right) to visualize the graph (or Riemann surface) of the complex inverse hyperbolic cosine function $w = \cosh^{-1} z$ constructed in a Cartesian coordinate 4-space, which is traditionally drawn in abstract form with polylines, as shown in Fig. 3. These are typical examples of the figurative approach, rather than topological abstraction, to mathematical visualization. As is well known, the inverse hyperbolic functions can be defined in terms of logarithms:

$$\cosh^{-1} z = \log \left(z + \sqrt{z^2 - 1} \right) = \log \left(z + \sqrt{z + 1}\sqrt{z - 1} \right). \tag{1}$$

This formula represents that the Riemann surface of the inverse hyperbolic cosine function can be constructed by piling up the two leaves fitted together (the Riemann surface of the square root; this is explained more in detail in Sect. 3) at both points $z = -1$ and $z = 1$ along the infinite spiral staircase (the Riemann surface for the logarithm). We can understand the meaning of this formula, which seems rather complicated at first sight, very easily and intuitively from the content of Fig. 2. Especially in mathematical visualization, we have demonstrated thus far that conceptualizing the fourth dimension in a figurative manner seems to be successfully justified by appropriately using a cross-section and a shadow of a mathematical object described in terms of Cartesian (or polar) 4-dimensional coordinates.

The procedures to obtain a projective line (or plane) by adding a point (or line) at infinity to the straight lines (or planes) that have been considered normal coordinate spaces are called *projectivization*. In our immersive mathematical environment described in the following sections, both procedures are intuitively visualized from the first-person perspective. For example, the Riemann sphere that appears in most primers of complex analysis (as shown in Fig. 4) is normally described as a real number line that is complexified first and then projectivized. In Sect. 5, we raise a question that has not been explored thus far: "What does the Riemann sphere's axis stand for?" We will show that the answer can be obtained only by observing the sphere from the inside by setting the viewpoint

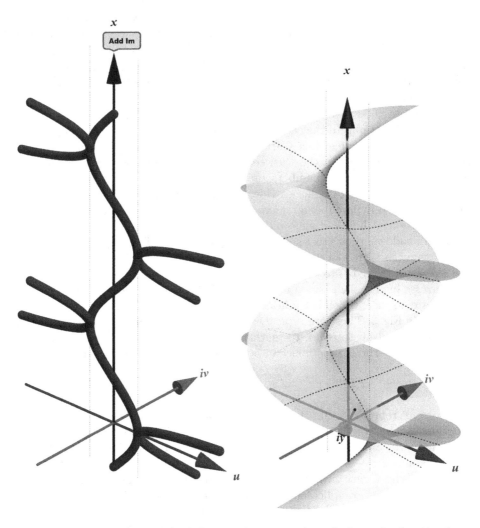

Fig. 2. Constructing the graph of the complex inverse hyperbolic cosine function in a Cartesian coordinate 4-space.

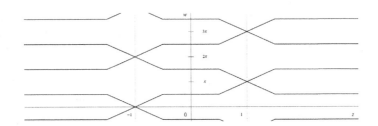

Fig. 3. The Riemann surface of $w = \cosh^{-1} z$ in traditional abstract form.

of the immersive environment to the point $(0,0)$, which is always undefined in projective geometry or, more specifically, in projective spaces based on homogeneous coordinates.

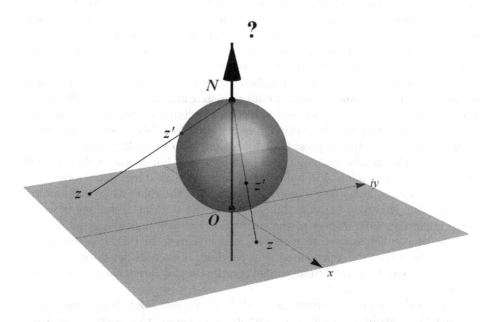

Fig. 4. Stereographic projection of a complex number z on point z' of the Riemann sphere.

2 High-Dimensional Data Visualizations

Dimension increase in the medium is an absolute necessity in dealing with high-dimensional data, both in a wide variety of information processing fields and in mathematical visualization. Many ideas and concepts in modern advanced mathematics are clearly defined using logic and mathematical expressions. Without the help of any visual elements, they naturally generalize to higher dimensions. However, dimensionality reduction is a significant problem in the data visualization field.

Data visualization methodology based on the mathematical ideas of coordinate systems began to emerge after René Descartes and Pierre de Fermat invented plane and solid analytic geometry in the 1630s. Since then, with the advent of computers, large amounts of data have been graphed, and visualization idioms have been developed mainly in the field of statistics.

A type of mathematical diagram that uses Cartesian coordinates to display values of two or more variables for a set of data is often referred to as *scatterplots*. For complex datasets, it is very difficult to manually identify a view that both maximizes the visibility of the relevant part of the data and minimizes occlusion.

Several attempts, such as the *matrix of scatterplots* [9,10], have been made to show all the pairwise variables on a single view with multiple 2-dimensional scattergrams in a matrix format. Classic idioms–the *line chart* and the *radar chart*–are also used to find a degree of correlation between multiple variables. Although these idioms do not have the function of dimensionality reduction themselves, many extensions have been studied, developed, and applied.

As complementary techniques for visualizing high-dimensional geometry, *heat maps* [11], *height maps*, and *domain coloring* [12] (a mathematical technique derived from heat mapping for visualizing complex functions) treat numerical values as colors and visualize spatial data as planar histograms appear. However, because the colors lack the natural perceptual ordering found in grayscale colormaps, we must to find a design that does not allow for misinterpretation. *Iconographic displays* [13] that express high dimensions with symbols were also devised to eliminate the cognitive ambiguity in data elements. Furthermore, *dimensional stacking* [13] was developed to reduce the number of high-dimensional axes. This is achieved by optimally and hierarchically dividing the chart while maintaining the visibility. However, because the concept of the stacking technique is sometimes ineffective with respect to the correlation between multiple variables, the concept of *parallel coordinate plots* [14] was invented. In this, a point in the n-dimensional space is represented as a polyline. If we consider dimensionality reduction, it can be argued that parallel coordinate plots are a natural extension of the classical line chart.

While each idiom has advantages and disadvantages, high-dimensional data visualizations evolved to allow a computer to achieve the desired goal with accuracy and precision. The reason is that information visualization deals with mostly discrete models, sometimes abstract and nonspatial data. Most mathematical visualization has to use continuous models such as expressions and function graphs and includes all visualization algorithms for utilizing continuous models of data. Even though the data values are discrete, it is considered that they can be understood as a constant phenomenon eventually. Descartes introduced Cartesian coordinates and extended the idea that "numbers can be thought of as corresponding to figures". From the late 19th century to the early 20th century, David Hilbert advocated "Foundations of Geometry", which was completely separated from the human senses for the sake of logical consistency. Extreme abstraction can be seen as a pioneer of today's axiomatic mathematics. Since then, however, has mathematics neglected the effort to give a figurative form to abstract concepts? On the other hand, in the field of recent information visualization, validating the effectiveness of a visualization design is important. "The design algorithms to instantiate the visualization idioms computationally are thought to be located in the center of the validation methods" [15]. We believe that it is vitally important to apply this concept to mathematical visualization.

3 Mathematical 4-Space Visualization Preliminaries

The aim of our research is primarily to point out that several issues exist in popular mathematical visualization, which no one has ever been claimed for a couple of decades. The answers to various problems of basic math consisting of elementary calculus and geometry faced by high-school and college students have already believed to be well-established, and there might not be any room for further study or discussion. But pitfalls are found almost everywhere in this area, especially when it comes to graphing functions to understand their behavior in a higher dimensional space that we can neither experience nor access directly. Although topological abstraction plays an exceptionally important role in modern mathematics, it can greatly reduce the amount of information to what is needed for the problem to be just identified and lead to a lot of misconceptions. Rather than such abstraction processes, we advocate that figurative approaches are often considered as prime solution candidates. As an example of the graphical representation of complex roots in solving a quadratic equation, it is already evident by a report written in 1945 that visualization is possible by increasing the dimension of the media from 2 to 3 (Figs. 5 and 6). Nonetheless, even now at more than 70 years since then, they are still uncommon.

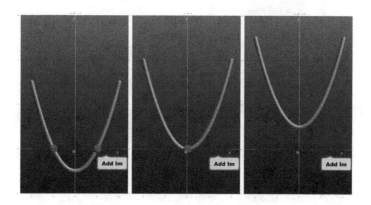

Fig. 5. Graphical representation of the roots of $w = z^2 - 1$, z^2 (double root), $z^2 + 1$ (no root).

We can reconstruct the exact form of some real functions from fragmentary facts while making an analytic continuation with their singularities. Can we visualize the overall behavior as a complex function by using graphs as in real functions? Unfortunately, few textbooks have answered this question directly. Lars Valerian Ahlfors' *Complex Analysis*, a classic masterpiece published in 1953, states that the conformal mapping associated with an analytic function affords an excellent visualization of the properties of the latter; it can well be compared with the visualization of a real function by its graph. Ahlfors seems to have given up on the attempt to obtain a graph of a complex function [16]. Is trying to visualize complex functions with the concept of graphs indeed hopeless?

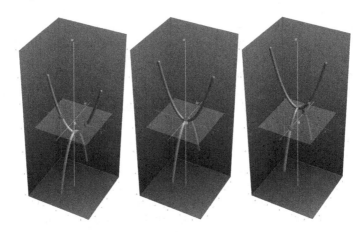

Fig. 6. Graphical representation of complex roots after an extra dimension is added.

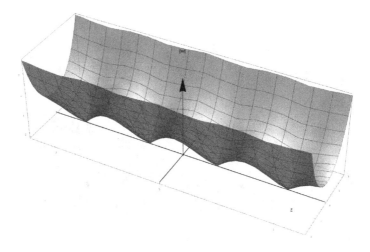

Fig. 7. Modular surface of $|\cos z|$.

As a method for visualizing complex functions, the *modular surface* [17] trial draws a graph with three axes, x, y, and $|w|$, by using the absolute value of the function value w. Figure 7, a graph of $|\cos z|$, can reasonably be realized in a 3-dimensional space. This method is generally useful. As seen in the figure, identifying the features of trigonometric functions that vibrate on the real axis and diverge in the imaginary axis direction is also possible. However, the absolute value is the distance from the origin; in other words, this method loses information on the argument of the function value. More specifically, the original zero point becomes a point where it cannot be differentiated; even multivalent functions are sometimes drawn as being monovalent, and care must be taken to ensure that the top graph is far away from the true graph in terms of topology. If we deal with a function that takes only real values from the beginning, therefore,

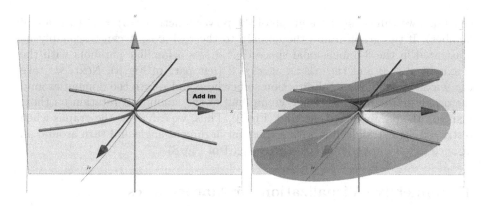

Fig. 8. Constructing the graph of the complex square root function around its singular point in terms of polar coordinates.

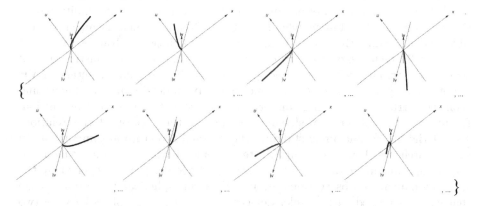

Fig. 9. Bracketing all the cross-sections along the argument direction of polar coordinates.

we must abandon the visualization of all the most complex analytic functions that are most familiar. The function defined by the open area of the complex plane is differentiable (regular), the Cauchy-Riemann equation is established, and the power series can be expanded (analytical) at each point of the domain. All these contractive conditions are thus proven equivalent. A beautiful theorem is unexpected in actual analysis, which is the starting point of a complex analysis. From this fact, $v = v(x, y) = 0$ for a real-valued function, so from the Cauchy-Riemann equation $u_x = v_y$, $u_y = -v_x$, $u_x = 0$, $u_y = 0$. It cannot be other than a constant function (Fig. 9).

Although it is somewhat difficult, the absolute value surface shown above has succeeded in qualitatively expressing the complex function by considering the map $|f| : \mathbf{R}^2 \to \mathbf{R}$, in which the range is made 1-dimensional. In contrast, from the graph of the so-called vector value function, a map $f|_A : \mathbf{R} \to \mathbf{R}^2$ restricted on a half line $\mathbf{A} = \{z = re^{i\theta}, \theta \text{ is constant}\}$ on the z plane or, more generally, we will construct a graph of the power function $f(z) = z^\alpha$ (α is a real number). If the graph limits the domain to the real axis, it can reasonably be illustrated in the 3-dimensional space that draws a familiar parabola with the u axis at $x \geq 0$ and the v axis at $x \leq 0$ (left part of Fig. 8). Next, we take the y axis in the horizontal direction somewhere and rotate the domain around the origin gradually on it, while giving half of the amount of rotation on the z plane on the parabolic axis around (Fig. 9). Thus, we can draw the graphs when restricted on a straight line with a constant argument angle and turn around all at once in a 3-dimensional space (right part of Fig. 8).

4   Immersive Visualization for Mathematics

As the general display is 2-dimensional, it is necessary to forcibly project 3-dimensional to 2-dimensional and make the pseudo 3-dimensional space recognized. In the visualization of our high-dimensional space, dimensionality reduction is indispensable, but we should maintain 3-dimensional as much as possible and maximize the use of human cognitive abilities. Advances in virtual reality (VR) technology users to experience a 3-dimensional realistic sense of geometric attributes with a stereoscopic feeling. The essential advantage of VR is not direct perception of three dimensions. In understanding 3-dimensional geometric structures, the most important property is the illusion of immersion. Eliminating the influence of shielding because of the anteroposterior relationship of the object and understanding the internal structure are important. In this research, we show that complete stereoscopic viewing with a head-mounted display and an understanding of 3-dimensional geometry are deepened by the effect of immersive observations, even with a full spherical panoramic image that can be handled with simpler devices. One typical example is locating two circular-points-at-infinity (Fig. 1). Although many books on geometry [18–20] have been introduced the points in the complex and infinitely distant territory, they have never been rendered so far.

5   True Colors of the Riemann Sphere

In geometry, the Riemann sphere is the prototypical example of a Riemann surface and is one of the simplest complex manifolds. In projective geometry, the sphere can be considered the complex projective line CP^1, the projective space of all complex lines in C^2. We raise a question that has not been explored thus far: "What does the Riemann sphere's axis stand for?" Is it one of the ranges (u or v) of the complex function $u + iv = f(x + iy)$? As the Riemann sphere is the domain of a function that is projectivized and then compactified, the range (codomain) has no relation in the first place. The solution candidate is the axis of "1" which we use when we define $(z, 1)$ as the homogeneous coordinates of z. In this case, however, the 1-dimensional axis is strange in that no imaginary axis exists. In addition, it is even stranger that something (usually the center or the

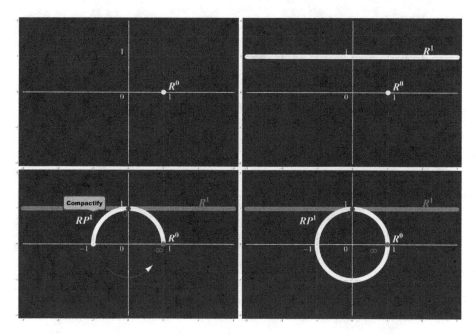

Fig. 10. Recurrence relation between the real projective spaces.

south pole of a sphere) is placed at the origin $(0,0)$, which is always undefined in projective geometry.

Many discussions and progressions have been omitted for simplicity, and the Riemann sphere is suddenly given as an abstracted conclusion from the topological point of view. Thus, we can confirm from the graphics and mathematical equations what the complex version of the recurrence relation between projective spaces $\boldsymbol{RP}^n = \boldsymbol{R}^n \cup \boldsymbol{RP}^{n-1}$ (see Figs. 10 and 11) means: $\boldsymbol{RP}^1 = \boldsymbol{R}^1 \cup \boldsymbol{R}^0$ means the complex projective line (i.e., a Riemann sphere) after complexification.

We will show that the answer can be obtained only by observing the sphere from the inside by setting the viewpoint of the immersive environment on the point $(0,0)$, which is always undefined in projective geometry or, more specifically, in projective spaces based on homogeneous coordinates (Figs. 12 and 13). These results have never been achieved with a traditional abstract and topological manner.

The Riemann sphere's axis expresses the real part r of the complex homogeneous coordinate t, such that $t = r + is$. Notice that we have to set the center of the sphere to the origin of the coordinate system.

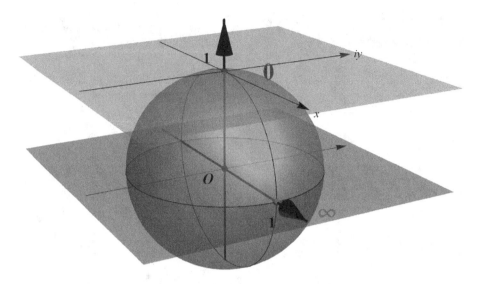

Fig. 11. True colors of the Riemann sphere represented in the cross-section of the complex projective plane.

Fig. 12. A complexified circle (gray) and the Riemann sphere (white).

Fig. 13. By moving the viewpoint of the immersive environment from the outside to the inside of the Riemann sphere, the texture on the complexified circle gradually matches the texture on the spherical surface (from left to right, top to bottom).

6   Conclusion

We can show that the understanding of 3D geometry will be deepened by the effect of immersive observations based on a figurative and non-topological visualization approach, rather than in an abstract way. By further expanding the above-mentioned research results, we plan to develop a visualization method applicable to CP^2 (complex projective plane).

Acknowledgment. The present work has been financially supported in part by MEXT KAKENHI Grant-in-Aid for Scientific Research on Innovative Areas No. 25120014.

References

1. Fisher, S.S., Humphries, J., McGreevy, M., Robinett, W.: The virtual environment display system. In: 1986 ACM Workshop on Interactive 3D Graphics (1987)
2. Taxén, G., Naeve, A.: CyberMath: exploring open issues in VR-based learning. In: SIGGRAPH 2001 Educators Program, SIGGRAPH 2001 Conference Abstracts and Applications, pp. 49–51 (2001)
3. Yeh, A., Nason, R.: VRMath: a 3D microworld for learning 3D geometry. In: Proceedings of World Conference on Educational Multimedia, Hypermedia and Telecommunications, Lugano, Switzerland (2004)
4. Kaufmann, H., Dünser, A.: Summary of usability evaluations of an educational augmented reality application. In: Shumaker, R. (ed.) ICVR 2007. LNCS, vol. 4563, pp. 660–669. Springer, Heidelberg (2007). https://doi.org/10.1007/978-3-540-73335-5_71

5. Trien Do, V., Lee, J.W.: Geometry education using augmented reality. Paper presented at Workshop 2: Mixed Reality Entertainment and Art (at ISMAR 2007), Nara, Japan (2007)

6. Miyazawa, A., Nakayama, M., Fujishiro, I.: A figurative and non-topological approach to mathematical visualization. In: Proceedings of 2018 International Conference on Cyberworlds, pp. 150–155. Nanyang Technological University, Singapore, October 2018. https://doi.org/10.1109/CW.2018.00036

7. Hinton, C.H.: The Fourth Dimension, 2nd edn. Ayer Co., Kessinger Press reprint (1912, orig. 1904), scanned version available online at the Internet Archive, https://archive.org/details/fourthdimension00hintarch. Accessed 21 Dec 2018

8. Kaku, M.: Hyperspace: A Scientific Odyssey Through Parallel Universes, Time Warps, and the 10th Dimension, pp. 68–74. Oxford University Press, Oxford (1995)

9. Andrews, D.F.: Plots of high-dimensional data. Biometrics **28**, 125–136 (1972). https://doi.org/10.2307/2528964

10. Chambers, J.M., Cleveland, W.S., Kleiner, B., Tukey, P.A.: Graphical Methods for Data Analysis. Wadsworth International Group, Belmont (1983). 978-0534980528

11. Pickett, R.M., Grinstein, G.G.: Iconographic displays for visualizing multidimensional data. IEEE Conf. Syst. Man Cybern. **514**, 519 (1988). https://doi.org/10.1109/ICSMC.1988.754351

12. Farris, F.A.: Visualizing complex-valued functions in the plane. http://www.maa.org/visualizing-complex-valued-functions-in-the-plane. Accessed 19 Dec 2017

13. Ward, M.O., LeBlanc, J., Tipnis, R.: N-land: a graphical tool for exploring N-dimensional data. In: Computer Graphics International Conference, p. 14 (1994)

14. Inselberg, A., Dimsdale, B.: Parallel coordinates for visualizing multidimensional geometry. In: Computer Graphics International, pp. 25–44 (1987). https://doi.org/10.1007/978-4-431-68057-4_3

15. Munzner, T.: Visualization Analysis and Design. A K Peters Visualization Series, p. 72. CRC Press, Boca Raton (2014). 978-1466508910

16. Ahlfors, L.V.: Elementary conformal mapping. In: Complex Analysis, an Introduction to the Theory of Analytic Functions of One Complex Variable, p. 89. McGraw-Hill, New York (1953). 978-1124111148

17. Needham, T.: Complex functions as transformations. In: Visual Complex Analysis, p. 56. Oxford University Press, Oxford (1997). 978–0198534471

18. Hilbert, D., Cohn-Vossen, S.: Geometry and the Imagination. AMS Chelsea Publishing, Providence (1999)

19. Richter-Gebert, J.: Perspectives on Projective Geometry. A Guided Tour Through Real and Complex Geometry, 1st edn. Springer, Heidelberg (2011). https://doi.org/10.1007/978-3-642-17286-1. Softcover reprint of the original

20. Maxwell, E.A.: The Methods of Plane Projective Geometry Based on the Use of General Homogenous Coordinates. Cambridge University Press, Cambridge (1946)

Fast 3D Scene Segmentation and Partial Object Retrieval Using Local Geometric Surface Features

Dimitrios Dimou[1(✉)] and Konstantinos Moustakas[2]

[1] Institute for Systems and Robotics, Instituto Superior Tecnico, Lisbon, Portugal
mijuomij@gmail.com
[2] Department of Electrical and Computer Engineering,
University of Patras, Patras, Greece
moustakas@upatras.gr

Abstract. Robotic vision and in particular 3D understanding has attracted intense research efforts the last few years due to its wide range of applications, such as robot-human interaction, augmented and virtual reality etc, and the introduction of low-cost 3D sensing devices. In this paper we explore one of the most popular problems encountered in 3D perception applications, namely the segmentation of a 3D scene and the retrieval of similar objects from a model database. We use a geometric approach for both the segmentation and the retrieval modules that enables us to develop a fast, low-memory footprint system without the use of large-scale annotated datasets. The system is based on the fast computation of surface normals and the encoding power of local geometric features. Our experiments demonstrate that such a complete 3D understanding framework is possible and advantages over other approaches as well as weaknesses are discussed.

Keywords: 3D computer vision · Geometric 3D segmentation · Partial 3D object retrieval

1 Introduction

Over the past years, various affordable and high-quality depth sensors (such as Microsoft's Kinect, Intel's RealSense etc.) have been made available to the public. This fact along with the creation and the wide availability of RGB-D datasets [9], has lead the research community to develop advanced methods and algorithms for the processing of 3D information.

The methods developed for 3D perception have found a variety of applications in a wide variety of settings, from real-world problems in robotics, such as robotic grasping [41], and robotized waste sorting [11], to autonomous driving systems [37], 3D reconstruction [40] and virtual and augmented reality. Although each field meets specific needs and constraints, the fundamental problems posed

© Springer-Verlag GmbH Germany, part of Springer Nature 2020
M. L. Gavrilova et al. (Eds.): Trans. on Comput. Sci. XXXVI, LNCS 12060, pp. 79–98, 2020.
https://doi.org/10.1007/978-3-662-61364-1_5

in every application is the segmentation of a three-dimensional scene to its containing elements, i.e. the objects that constitute it, and the classification of each individual object or the retrieval of similar ones. Additionally, in most of the applications, the time and memory requirements are of critical importance as the deployment devices are usually embedded processors that have limited processing and memory capacity, thus making it prohibited to employ vast architectures such as deep neural networks that contain million of parameters.

Furthermore, although there has been a considerable amount of research in 3D scene segmentation and partial 3D object retrieval, most of the previous work tackles the problems separately, thus not addressing problems encountered in between stages of the pipeline. The introduced steps between the segmentation and the retrieval module are of critical importance in order to pass from the simple and noisy representation of the point cloud taken directly from the depth sensor to a more suitable representation for the extraction of surface descriptors.

To this end, we present a complete, real-time framework for the segmentation and retrieval of partial objects present in 3D scenes. A preliminary version of this work has been reported in [6], where we presented a brief overview of the proposed architecture. In this work, we present a more in depth analysis of the proposed methods, incremental computational improvements as well as extended experimental results to support our design choices. Our approach is based on the fast geometric segmentation of the depth scene into continuous surface patches, the pre-processing of each surface patch in order to reduce noise and computational complexity for the next steps and finally the use of geometrical and statistical features for the computation of similarity distances between our partial patches and the complete models from the database. Our main contributions are:

- Additional filtering steps for the edge detection at the segmentation stage.
- Introduction of normalizing stages, and the use of kd-trees for more efficient distance computation.
- The introduction of a sequence of steps for the pre-processing of the partial patches.

The remaining of this paper is organized as follows: The next section presents the current state-of-the-art methods for the tasks of 3D scene segmentation and 3D object retrieval and highlights the pros and cons of each method. Section 3 provides a complete overview of the proposed framework along with examples and finally in Sect. 4 we evaluate our approach and compare it with other recent methods.

2   Related Work

Most of the previous work on 3D scene understanding focuses either on the scene segmentation task or the object retrieval task. Although, there has been some work addressing both problems in a unified framework and is discussed at the end of this section.

2.1   Segmentation

Scene segmentation approaches usually employ geometric features and common classification schemes in order to successfully label each pixel. The recent advances in 2D computer vision using deep learning techniques and the availability of large annotated datasets has made it possible for the development of deep learning models for 3D semantic segmentation, which constitute the state-of-the-art in the field.

Traditional Methods. Most traditional methods can be roughly divided in geometry based methods and classification based methods.

Geometric methods can be based on edges, regions or pre-existing models. In [14], they use a fast triangulation scheme to compute approximate local surface normals and finally a region growing based segmentation to label each point. In [41], they compute directly the surface normals from the depth image and use them to detect edges, then using a simple region growing algorithm they group pixels into surface patches, in order to combine the patches into object hypotheses they determine a directed, weighted graph, modeling the adjacent patches and subsequently they find the most probable segmentation into object regions in a greedy manner. In [7], they take advantage of the 3D symmetries detected on the scene, which are used to find consistent segmentations. Finally, in [34], they build a global segmentation map by propagating and merging segments extracted from each depth map, via a simultaneous localization and mapping (SLAM) reconstruction algorithm.

In most classification based methods, 3D features are used and common classifiers to cluster points into regions. In example, in [12] they use eigenvalues and eigenvectors of covariance tensors in order to compute a point feature and a random forest classifier to classify each point. In [39], they propose a sequential parsing procedure that learns the spatial relationships of objects. Finally in [35] they combine multiple segmentation and post-processing methods to achieve useful point cloud segmentations.

Deep Learning Methods. Deep learning approaches operate mainly on point clouds and use various representations in order to apply 3D convolutions. In [15], the authors propose a 3D Convolutional Neural Network (CNN) which takes as input a voxelized point cloud and produces exactly one label for each voxel. In [4], they propose a framework which applies CNNs on multiple 2D snapshots of the point cloud and finally project the label predictions back in the 3D space. In [36], after extracting features using the PointNet framework they introduce a Similarity Group Proposal Network that takes advantage of a similarity matrix between pair of points in embedded feature space, thus producing an accurate 3D instance-aware semantic segmentation on the point cloud. In [8], they built upon the PointNet architecture and propose two extensions to increase the context of the network by considering a group of blocks simultaneously.

2.2 Retrieval

Methods addressing partial 3D object retrieval can be divided based on the dimensionality of the representation data [28] to 2D (view-based) representations and 3D (model-based) representations. For a more comprehensive review of recent object retrieval methods the reader should see [25].

View-Based Methods. One common approach among the view-based methods is the query-by-range-image approach. In [17] salient features are extracted on the query image and the 3D models and the matching is done via estimating the dissimilarity within the protrusion filed, whereas in [31] the same salient features are used to perform a hierarchical search in the parameter space for an optimum solution. Papadakis et al. in [19] introduced a novel 3D shape descriptor based on a set of panoramic views of a 3D object. The descriptor is obtained after the object is projected onto three perpendicular cylinders and for each projection they compute the corresponding 2D Discrete Fourier Transform as well as 2D Discrete Wavelet Transform. Finally the similarity between two 3D objects is measured by the distance between of the descriptors. In [27], they use the same method for extracting panoramic views, but for each view they compute the enhanced DSIFT descriptor and finally use a Bag-of-Visual-Words model for matching the 3D models. In [28], a Convolutional Neural Network is used, trained on an augmented view of the extracted panoramic representation, for classification and retrieval of 3D models. In [29], the authors design a CNN specifically for learning deep representations directly from panoramic views. CNNs are also used directly in [30], after converting the 3D shapes into geometry images which encode local properties of shape surfaces such as principal curvatures. Finally, in [13], the Multi-view Convolutional Neural Network framework [32] is used, but instead of the traditional classification loss, they propose a triplet-centered loss which has a center for each class and requires that the distances between samples and centers from the same class are closer than those from different classes.

Model-Based Methods. 3D representations are usually based on extracting local or global model descriptors and formulating similarity distances for matching. One very popular shape descriptor is the Heat Kernel Signature (HKS) [33], in [5] a scale-invariant version is introduced based on a logarithmically sampled scale-space which is later used in the bag-of-features framework for shape retrieval. Another popular choice is descriptors based on the Point Feature Histograms (PFH) [23], in [22] Fast Point Feature Histogram (FPFH) is introduced which reduces the computational complexity and is used for 3D registration and in [26] an extended version of the descriptor is used for measuring local similarity along with the Fisher Encoding for measuring global similarity.

2.3 Combined

Finally, there are certain approaches that aim at a simultaneous segmentation and recognition of 3D objects. In [16], they try to recognize and localize queried objects in depth images by maximizing the use of the visibility context.

They present a novel point pair feature that is based on the visibility context and use to recognize and estimate the pose of the queried objects. In [21], they design a new type of neural network that takes as input unprocessed point clouds and can be trained for a number of 3D recognition tasks, such as object classification, semantic segmentation etc.

3   Framework Overview

This section outlines the overall structure of our pipeline. It is compiled of three main parts: the segmentation of the depth image, the pre-processing of each partial object view and the retrieval of similar 3D models from a database. The end goal of our system is to retrieve 3D objects similar to the ones pictured in our 3D scene.

To this end, we employ an edge-based segmentation algorithm, in order to find continuous and compact surface patches in the scene. This geometric approach exhibits low computational complexity, while produces reliable surface hypothesis. Though it does not aim to find semantically meaningful objects. Then each individual patch is pre-processed and fed into the retrieval system, where geometric and statistical features are mixed and geometrically similar 3D objects are retrieved from the database.

Both methods are based on primitive geometric features such as normals, that can be computed quickly. Even the statistical features in the retrieval module are extracted from the relations of geometric features. This allows faster processing times and a lower memory footprint as features, such as the normals, computed in the segmentation module can be reused for the retrieval stage, which is an important advantage of the combination of these specific methods. In addition, the processing of the target objects can be done offline thus leaving the processing of the query patch and the distance computation during test time. In our experiments, we focus mainly on tabletop scenes as they constitute a more popular framework in robotic manipulation and human interaction tasks.

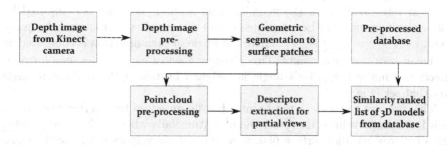

Fig. 1. Example of the pre-processing pipeline. From left to right we see an initial noisy point cloud patch, the point cloud with the removed outliers, the noisy surface patch and finally the smoothed surface patch.

3.1 Geometric Segmentation

As the computational complexity of the used methods is of great importance. As one of our main concerns for our framework is the computational complexity of the used methods, we decided to follow the work of [41], operating directly on the raw depth maps, ignoring color, as the recently developed sensing devices provide high-quality depth information.

The first step is to estimate the surface normals for each point (pixel) of the depth map, as it is the principal geometric feature upon which the edge detection algorithm is based. The normals are approximated with the normal vector of the isosceles triangle plane formed by three symmetrical points, in the $n \times n$ neighborhood around the central point, where n is a system hyperparameter. The calculation is based on the cross product rule between the two vectors formed by the triangles vertices (see Fig. 2).

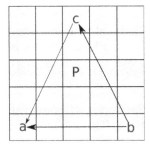

Fig. 2. Example of points (pixels) used for the estimation of the normal vector in a 3×3 neighborhood around a point p_q.

The next step is to determine which image points belong to a surface edge. In order to achieve that an angle value is assigned to each point through a max-mean scheme. First, given a query point p_q the angles in each direction (four cardinal and four intercardinal) in the neighborhood around the point is computed by taking the inner product between the normal vectors of each point and the query point, i.e. $\theta = cos^{-1}(\frac{p_q p_t}{\|p_q\|\|p_t\|})$. Then, the angle in each direction is computed as the mean over the used points. This gives us a set of eight angles, one for each direction. Finally the point's angle is assigned by taking the maximum angle over that set (Fig. 1).

This way we can detect the points that lie on edges by filtering the points with large angle values. However, the algorithm fails when we have very sharp edges between parallel surface planes, as the normal vectors will have the same direction and the angle value for the points will not be sufficient large. On this account, we add another filter that detects sudden jumps of depth values along the eight directions used for the angle computation. Along with the surface normal angles in each direction we compute the mean depth difference among the considered pairs of points.

The above manipulation results in an image that each pixel is associated with two values one for angle and one for the depth transition. This image is then binarized using a thresholding value for the angles and for the depth differences, and strong edges become easily distinguishable from smoothly curved surfaces. Consequently, the flood fill algorithm is applied in the edge image in order to detect the continuous surfaces that lie inside strong edges. Finally the points that are part of the detected edges are assigned to the neighboring surfaces which has the smallest Euclidean distance, this is achieved by searching the neighborhood around a point in a breadth-first manner. Results of the segmentation steps from a sample image from the used dataset can be seen in Fig. 3. At the end of this processing step we obtain a set of different surfaces on the depth image, that are represented with color labels, which are then reconstructed to a set of point clouds, using the intrinsic device's parameters, and fed to the next processing module. Note that we aim at a geometric surface segmentation and not in a semantic segmentation, thus we do not follow the aforementioned methods, that have additional processing steps in order to determine whether the different surface segments could belong to the same object.

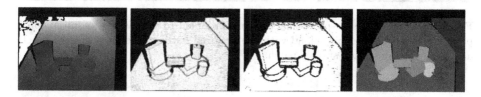

Fig. 3. Results for the various segmentation steps. From left to right we see the original depth image, edge image (value of angle and depth difference represented as pixel intensity), binarized image, final segmentation result.

3.2 Patch Pre-processing

In this stage the detected surfaces are pre-processed in order to be fed into the retrieval system. The aim is to alleviate some of the noise introduced from the depth sensor and the segmentation step, and output a better representation for the similarity assessment. The first step is to cleanse the partial surface by removing any outliers introduced from the measurement or the segmentation. To this end, we apply the statistical outlier filtering method introduced in [24], which is based on the statistical analysis of each points distance with its neighbors. More specifically, the remaining points belong to the following set:

$$\mathcal{P}^* = \{p_q \in \mathcal{P} | (\mu_k - \alpha \cdot \sigma_k) \le \overline{d} \le (\mu_k + \alpha \cdot \sigma_k)\}$$

where $\mathcal{P}$ is the initial partial point cloud, μ_k, σ_k is the mean and standard deviation of the distances between the query point p_q, which are calculated in the neighborhood around the point, with the number of neighbors to consider as a hyperparameter and α is a hyperparameter denoting the resilience to outliers.

In order to avoid problems with the scales of the objects and having to apply different hyperparameters in later steps, we subsequently normalized the patch, i.e. we perform a translation so as it's centroid lies at the origin of the axes and a scaling transformation so all points are contained in the unit sphere.

Finally, a bilateral [10] and a multilateral [14] filter is applied alternately in order to smooth the surface patch. This step acts as a high frequency filter, removing noise introduced from the measurement device, while preserving edges. Note that these filters require surface normal information so a first approximation is used by applying the PCA method from [24]. And then we estimate again the normals from the smooth point cloud patch to obtain the final set of points and normals that will be used in the similarity assessment step.

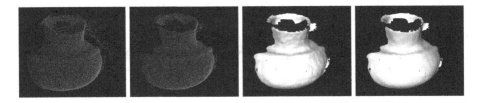

Fig. 4. Example of the pre-processing pipeline. From left to right we see an initial noisy point cloud patch, the point cloud with the removed outliers, the noisy surface patch and finally the smoothed surface patch.

3.3   Descriptor Extraction and Matching

In this section, we describe the pipeline we use in order to measure the similarity between an partial query patch and the complete target models from our database. To this end we use two similarity measures, one for the local similarity between the points of the query and the target object, that is based on local geometric surface features, and one for the global similarity of the models, that is based on trained statistical features. In order to reduce the computational costs of these step, we first downsample the input point clouds using a voxel grid filter to obtain a quasi-uniform sampled point cloud, with the sampling density as a parameter. The normal vectors for each point are approximated as a weighted average of the normals of the points inside each voxel.

For the description of a point and its the local neighborhood we use the we use the differential fast point feature histogram signature (dFPFH), introduced in [26], which extends the FPFH [22] by capturing local geometric transitions in a ribbon shaped neighborhood around each point. The FPFH acts as a local geometry descriptor that encodes a more generalized notion of curvature using a multi-dimensional histogram of values. We store these descriptors in a k-d tree [3], in order to achieve faster nearest neighbor searches during the similarity assessment stage (Fig. 4).

For the global description of the point cloud the Fisher Vector encoding is used which provides a global feature vector for the models. The encoding starts by learning a Gaussian Mixture Model (GMM) θ, on the dFPFH signatures, by means of an expectation-maximization algorithm. And then for each Gaussian component the average first and second order differences between the point signatures (dFPFH) and the centers of the GMM are computed. The final Fisher Vector is derived from the concatenation of these difference vectors. Note that for the partial queries, their Fisher Vectors are encoded using their dFPFH signatures and the GMMs of each of the target models, in order to describe the probability that the dFPFH signatures were generated by the targets GMM.

Given these two descriptions for each object, namely the k-d tree with the dFPFH signatures and the Fisher Vector, we define the overall similarity between a partial query Q and a complete object T, similarly to [26], as a weighted average between two distances: the local similarity and the global similarity. The local similarity is calculated by averaging the k minimum L_2 distances between pairs of dFPFH signatures and the global similarity by the weighted sum of L_2 distances between the most similar pairs of Fisher Vectors. After the similarity between the partial query and each 3D model from the database is calculated, the models are sorted in increasing order and we obtain a similarity ranked list of the database, from which the N most similar models are retrieved.

3.4 Database Processing

The database that contains the target 3D models in processed off-line and the descriptors are saved in a binary file. This way we can store only the required feature vectors for each target and not all the information about the models, like point positions, normal vectors, etc. which decreases the memory footprint of our framework, e.g. for a model of 12000 points, it will be downsampled to around 200 points, and after processing we will have a matrix of $200 \times 66 + 1330$ features saved in our database. In addition, note that the annexation of models in the database becomes trivial as we only need to pre-process the new model and save its descriptors in the database.

The processing of the database models involves the calculation of the normal vectors, that can be done using the well-known computer graphics method, as the models are usually represented as triangular meshes, which has lower complexity than using the PCA estimation. Then the models are normalized and downsampled with the voxel grid filter and finally, the dFPFH signatures are extracted and the Fisher Vector encoding is employed, note that the estimated GMM of each object is saved as well to be used in the encoding of the partial queries FV encoding step.

4 Experiments and Discussion

4.1 Experimental Setup

For the evaluation of our system we used two publicly available datasets. For the evaluation of the segmentation module we used the Object Segmentation

Database (OSD) [1], which provides RGB-D tabletop scenes with an increasing level of complexity, from simple scenes picturing relatively few objects to highly cluttered scenes with multiple stacked objects. The OSD is commonly used as a benchmark for robotic vision tasks. As our main requirement from the segmentation module is to find surface patches that belong to objects and not segment whole objects, as it would be much more computationally expensive, we did not perform a quantitative evaluation but rather a qualitative one as it can be seen in Fig. 3.

For the retrieval module we used the Virtual Hampson Museum collection [2], which consists of 3D pottery models and has been used as a benchmark in several state-of-the-art methods and in the SHREC 2016 competition on partial object retrieval and it consists of 383 models classified to 6 geometrical classes, examples of objects present in the different classes can be seen in Fig. 5.

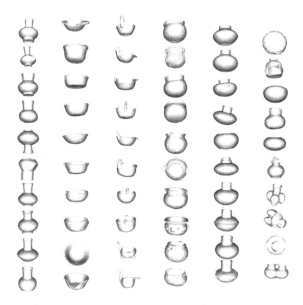

Fig. 5. Examples for the different classes present in the Hampson dataset.

Our experimental evaluation is based on precision-recall (P-R) curves, which is a commonly used measure in information retrieval tasks. We used the query dataset from the SHREC'16 Track on Partial Shape Queries for 3D Object Retrieval [20] which consists of three distinct query sets. One of artificial queries with 25% and 40% partiality, one of real high quality queries, obtained with the smartSCAN Breuckmann scanner and real low quality queries obtained with a Microsoft Kinect V2 sensor. In Fig. 6, we see the number of objects from each class present in the target dataset, as well as the three query datasets.

Our retrieval algorithm uses a Gaussian Mixture Model, trained with the Expectation Maximization algorithm, for the computation of the global

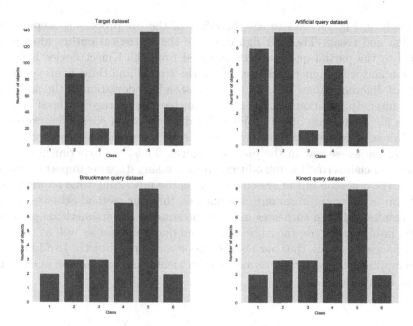

Fig. 6. Histograms of the number of objects for each class in the target and query datasets.

similarity between two objects, this introduces some stochasticity in the final computed similarity distance. In order to get a more accurate depiction of the results we run our algorithm with several initial random seeds and aggregate the final results[1].

Figures 7, 8, 9 present the quantitative results for each query dataset along with the corresponding running times. We can easily observe that the sampling parameter (voxel size) plays an important role in the retrieval performance of our system and its efficiency. More specifically, although we get comparable results with voxel sizes $0.1, 0.2$ the corresponding running time a voxel size 0.1 is almost three times more than the running time for voxel size 0.2 which makes it a forbidden option for real-time systems. Another noticeable result is that, although a voxel size of 0.1 improves the retrieval performance for small recall values, a voxel size of 0.2 decreases slower the precision of the system.

In Figs. 10, 11, 12 we see the obtained precision recall curves for each individual class, along with the mean performance over all classes. As we can see, for the classes 3 and 6 we achieve very low retrieval performance, which can be easily explained if we look at the objects that these classes include. Class 3 contains objects that their overall geometry is very similar with the ones from class 2 and class 6 contains objects that could not be classified to another class but do not share similar geometric properties.

---

[1] Our source code will be available on Github upon publication.

Finally in Fig. 13, we can see the effect of the pre-processing steps on the precision and recall. The first figure, where the statistical outlier filter is not applied to the partial query object obtained from the Kinect device has lower precision and with higher variance. While the second and third figure, show the effects of different values (0.5, 0.1), for the standard deviation of the statistical outlier filter. In addition, in Fig. 14, we can see the average retrieval times in seconds for a database of 100 objects, for different voxel sizes. The processing time decreases exponentially with the increase of the voxel size, while preserving the precision as we saw in the previous figures. Thus, the experimental results support our claims that the introduced pre-processing steps are important factors in both the precision of the system as well as the processing time requirements.

It can be observed, from our experiments, that our method achieves comparable results and even surpasses most of the state-of-the-art methods presented in [20], note that we use the same target and query dataset as well as the same performance measures, thus our results can be compared directly. In addition, its low computational performance makes it an ideal candidate for use in efficiency critical applications.

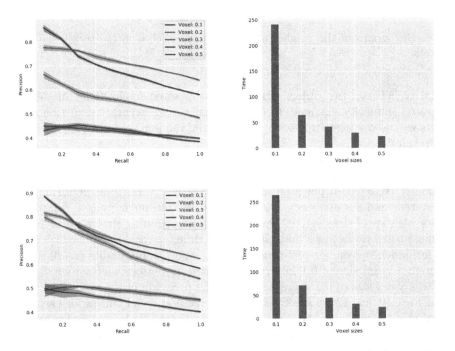

Fig. 7. Results on the Artificial query set, with 25% (top) and 40% (bottom), for different voxel sizes and the corresponding running times.

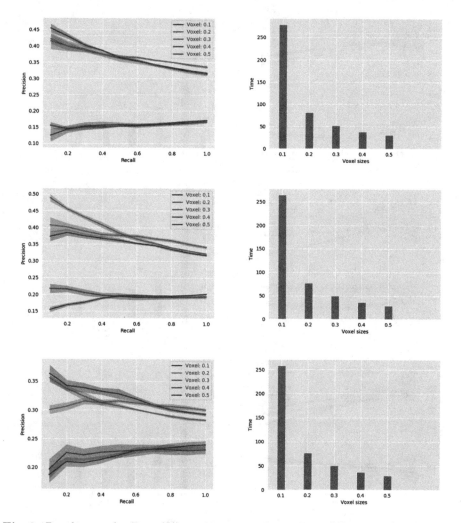

Fig. 8. Results on the Breuckmann query set, from three different viewpoints, for different voxel sizes and the corresponding running times.

4.2 Limitations and Future Work

The introduced understanding system provides state-of-the-art results in a variety of complex environments. However, the results remain far from perfect in absolute numbers. Both, the segmentation and the retrieval module fail to handle successfully different scenes under certain circumstances.

In particular, the segmentation algorithm fails in the presence of open curved objects, of objects split by complete occlusion and aligned object surfaces. The integration of color and texture information could solve some of the presented problems. In addition, it is rarely the case that the processing is done on individual frames. Usually the information comes from a video capturing device,

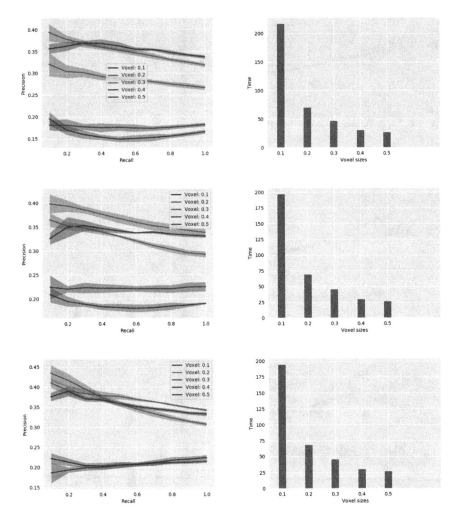

Fig. 9. Results on the Kinect query set, from three different viewpoints, for different voxel sizes and the corresponding running times.

thus making it possible to integrate temporal information as well, such as in [38], where a system that reconstructs high quality globally consistent point clouds is developed based on KinectFusion [18]. Such information could significantly improve the retrieval stage as well, as smoother and higher quality surface representation would generate more accurate descriptors. Another disadvantage with the proposed approach is that it needs uniformly sampled models in order to extract meaningful descriptors, for example it fails to retrieve CAD models as their representations usually have far less vertices than regular triangular meshes, in that case an additional pre-processing step would be needed to perform a quasi-uniform sampling over the surface. Finally, another interesting

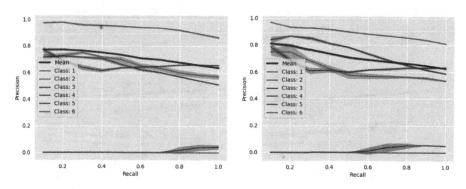

Fig. 10. Per class results on the Artificial query set [2].

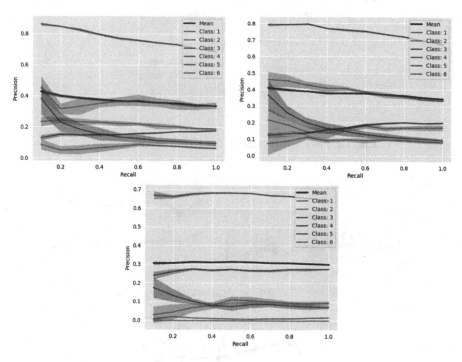

Fig. 11. Per class results on the Breuckmann query set [2].

direction for future work would be to use a more hierarchical structure in order to avoid having to sort the whole database of 3D models, but instead select the N most similar ones using for example a suitable global distance, and then use local distances in order to sort this subset of models.

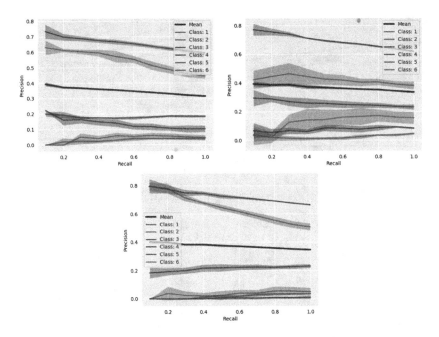

Fig. 12. Per class results on the Kinect query set [2].

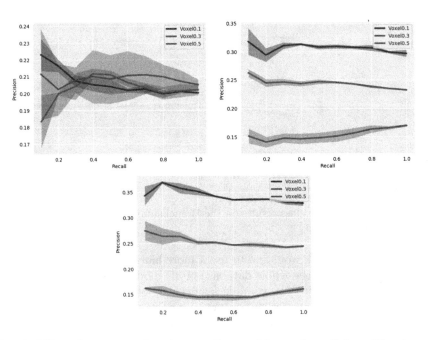

Fig. 13. Effect of pre-processing steps on the precision and recall for a Kinect query set.

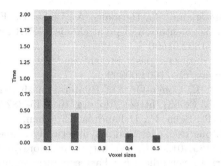

Fig. 14. Retrieval times for different voxel sizes in a database of 100 objects.

5   Conclusion

In contrast to previous works, that tackle the tasks posed in the problem of 3D scene understanding separately, we have presented a complete framework, that addresses the whole process in real world scenarios, from the capturing of a depth image, using a commodity depth sensor to the fast segmentation of the 3D scene and the retrieval of similar 3D models. Our system, which consists of three distinct parts, namely the segmentation module, the patch pre-processing module and the retrieval module, achieves state-of-the-art results on the tasks of 3D object segmentation and 3D partial object retrieval, although these tasks remain open problems for the research community, as they continue to be very challenging.

References

1. https://www.acin.tuwien.ac.at/en/vision-for-robotics/software-tools/osd/
2. http://hampson.cast.uark.edu/
3. Bentley, J.L.: Multidimensional binary search trees used for associative searching. Commun. ACM **18**(9), 509–517 (1975). https://doi.org/10.1145/361002.361007
4. Boulch, A., Saux, B.L., Audebert, N.: Unstructured point cloud semantic labeling using deep segmentation networks. In: Pratikakis, I., Dupont, F., Ovsjanikov, M. (eds.) Eurographics Workshop on 3D Object Retrieval. The Eurographics Association (2017). https://doi.org/10.2312/3dor.20171047
5. Bronstein, M.M., Kokkinos, I.: Scale-invariant heat kernel signatures for non-rigid shape recognition. In: 2010 IEEE Computer Society Conference on Computer Vision and Pattern Recognition, pp. 1704–1711, June 2010. https://doi.org/10.1109/CVPR.2010.5539838
6. Dimou, D., Moustakas, K.: A framework for 3D object segmentation and retrieval using local geometric surface features. In: International Conference CyberWorlds, October 2018
7. Ecins, A., Fermüller, C., Aloimonos, Y.: Cluttered scene segmentation using the symmetry constraint. In: 2016 IEEE International Conference on Robotics and Automation (ICRA), pp. 2271–2278, May 2016. https://doi.org/10.1109/ICRA.2016.7487376

8. Engelmann, F., Kontogianni, T., Hermans, A., Leibe, B.: Exploring spatial context for 3D semantic segmentation of point clouds. CoRR abs/1802.01500 (2018). http://arxiv.org/abs/1802.01500
9. Firman, M.: RGBD datasets: past, present and future. CoRR abs/1604.00999 (2016). http://arxiv.org/abs/1604.00999
10. Fleishman, S., Drori, I., Cohen-Or, D.: Bilateral mesh denoising. ACM Trans. Graph. **22**(3), 950–953 (2003). https://doi.org/10.1145/882262.882368
11. Grard, M., Brégier, R., Sella, F., Dellandréa, E., Chen, L.: Object segmentation in depth maps with one user click and a synthetically trained fully convolutional network. CoRR abs/1801.01281 (2018). http://arxiv.org/abs/1801.01281
12. Hackel, T., Wegner, J., Schindler, K.: Fast semantic segmentation of 3D point clouds with strongly varying density. ISPRS Ann. Photogramm. Remote Sens. Spat. Inf. Sci. **III–3**, 177–184 (2016)
13. He, X., Zhou, Y., Zhou, Z., Bai, S., Bai, X.: Triplet-center loss for multi-view 3D object retrieval (2018)
14. Holz, D., Behnke, S.: Approximate triangulation and region growing for efficient segmentation and smoothing of range images. Robot. Auton. Syst. **62**(9), 1282–1293 (2014). https://doi.org/10.1016/j.robot.2014.03.013
15. Huang, J., You, S.: Point cloud labeling using 3D convolutional neural network. In: 2016 23rd International Conference on Pattern Recognition (ICPR), pp. 2670–2675, December 2016. https://doi.org/10.1109/ICPR.2016.7900038
16. Kim, E., Medioni, G.: 3D object recognition in range images using visibility context. In: 2011 IEEE/RSJ International Conference on Intelligent Robots and Systems, pp. 3800–3807, September 2011. https://doi.org/10.1109/IROS.2011.6094527
17. Moustakas, K., Stavropoulos, G., Tzovaras, D.: Protrusion fields for 3D model search and retrieval based on range image queries. In: Bebis, G., et al. (eds.) ISVC 2012. LNCS, vol. 7431, pp. 610–619. Springer, Heidelberg (2012). https://doi.org/10.1007/978-3-642-33179-4_58
18. Newcombe, R.A.: KinectFusion: real-time dense surface mapping and tracking. In: 2011 10th IEEE International Symposium on Mixed and Augmented Reality, pp. 127–136, October 2011. https://doi.org/10.1109/ISMAR.2011.6092378
19. Papadakis, P., Pratikakis, I., Theoharis, T., Perantonis, S.: PANORAMA: a 3D shape descriptor based on panoramic views for unsupervised 3D object retrieval. Int. J. Comput. Vis. **89**(2), 177–192 (2010). https://doi.org/10.1007/s11263-009-0281-6
20. Pratikakis, I., et al.: SHREC'16 track: partial shape queries for 3D object retrieval (2016)
21. Qi, C.R., Su, H., Mo, K., Guibas, L.J.: PointNet: deep learning on point sets for 3D classification and segmentation. CoRR abs/1612.00593 (2016). http://arxiv.org/abs/1612.00593
22. Rusu, R.B., Blodow, N., Beetz, M.: Fast point feature histograms (FPFH) for 3D registration. In: 2009 IEEE International Conference on Robotics and Automation, pp. 3212–3217, May 2009. https://doi.org/10.1109/ROBOT.2009.5152473
23. Rusu, R., Marton, Z., Blodow, N., Beetz, M.: Persistent point feature histograms for 3D point clouds, vol. 16, January 2008
24. Rusu, R.B., Marton, Z.C., Blodow, N., Dolha, M., Beetz, M.: Towards 3D point cloud based object maps for household environments. Robot. Auton. Syst. **56**(11), 927–941 (2008). https://doi.org/10.1016/j.robot.2008.08.005

25. Savelonas, M.A., Pratikakis, I., Sfikas, K.: An overview of partial 3D object retrieval methodologies. Multimedia Tools Appl. **74**(24), 11783–11808 (2014). https://doi. org/10.1007/s11042-014-2267-9
26. Savelonas, M.A., Pratikakis, I., Sfikas, K.: Fisher encoding of differential fast point feature histograms for partial 3D object retrieval. Pattern Recogn. **55**, 114–124 (2016). https://doi.org/10.1016/j.patcog.2016.02.003. http://www.sciencedirect.com/science/article/pii/S0031320316000595
27. Sfikas, K., Pratikakis, I., Koutsoudis, A., Savelonas, M., Theoharis, T.: Partial matching of 3D cultural heritage objects using panoramic views. Multimedia Tools Appl. **75**(7), 3693–3707 (2014). https://doi.org/10.1007/s11042-014-2069-0
28. Sfikas, K., Theoharis, T., Pratikakis, I.: Exploiting the PANORAMA representation for convolutional neural network classification and retrieval, April 2017
29. Shi, B., Bai, S., Zhou, Z., Bai, X.: DeepPano: deep panoramic representation for 3-D shape recognition. IEEE Signal Process. Lett. **22**, 2339–2343 (2015)
30. Sinha, A., Bai, J., Ramani, K.: Deep learning 3D shape surfaces using geometry images. In: Leibe, B., Matas, J., Sebe, N., Welling, M. (eds.) ECCV 2016. LNCS, vol. 9910, pp. 223–240. Springer, Cham (2016). https://doi.org/10.1007/978-3-319-46466-4_14
31. Stavropoulos, G., Moschonas, P., Moustakas, K., Tzovaras, D., Strintzis, M.G.: 3-D model search and retrieval from range images using salient features. IEEE Trans. Multimedia **12**(7), 692–704 (2010). https://doi.org/10.1109/TMM.2010.2053023
32. Su, H., Maji, S., Kalogerakis, E., Learned-Miller, E.G.: Multi-view convolutional neural networks for 3D shape recognition. CoRR abs/1505.00880 (2015). http://arxiv.org/abs/1505.00880
33. Sun, J., Ovsjanikov, M., Guibas, L.: A concise and provably informative multi-scale signature based on heat diffusion. In: Proceedings of the Symposium on Geometry Processing, SGP 2009, pp. 1383–1392. Eurographics Association, Aire-la-Ville (2009). http://dl.acm.org/citation.cfm?id=1735603.1735621
34. Tateno, K., Tombari, F., Navab, N.: Real-time and scalable incremental segmentation on dense slam. In: 2015 IEEE/RSJ International Conference on Intelligent Robots and Systems (IROS), pp. 4465–4472, September 2015. https://doi.org/10.1109/IROS.2015.7354011
35. Vosselman, G.: Point cloud segmentation for urban scene classification. Int. Soc. Photogramm. Remote Sens. (ISPRS) **1**, 257–262 (2013). https://doi.org/10.5194/isprsarchives-XL-7-W2-257-2013
36. Wang, W., Yu, R., Huang, Q., Neumann, U.: SGPN: similarity group proposal network for 3D point cloud instance segmentation. CoRR abs/1711.08588 (2017). http://arxiv.org/abs/1711.08588
37. Wang, Y., Shi, T., Yun, P., Tai, L., Liu, M.: PointSeg: real-time semantic segmentation based on 3D LiDAR point cloud. ArXiv e-prints, July 2018
38. Whelan, T., Kaess, M., Johannsson, H., Fallon, M., Leonard, J.J., McDonald, J.: Real-time large-scale dense RGB-D SLAM with volumetric fusion. Int. J. Robot. Res. **34**(4–5), 598–626 (2015). https://doi.org/10.1177/0278364914551008
39. Xiong, X., Munoz, D., Bagnell, J.A., Hebert, M.: 3-D scene analysis via sequenced predictions over points and regions. In: 2011 IEEE International Conference on Robotics and Automation, pp. 2609–2616, May 2011. https://doi.org/10.1109/ICRA.2011.5980125

40. Yücer, K., Sorkine-Hornung, A., Wang, O., Sorkine-Hornung, O.: Efficient 3D object segmentation from densely sampled light fields with applications to 3D reconstruction. ACM Trans. Graph. **35**(3), 22:1–22:15 (2016). https://doi.org/10.1145/2876504

41. Ückermann, A., Haschke, R., Ritter, H.: Real-time 3D segmentation of cluttered scenes for robot grasping. In: 2012 12th IEEE-RAS International Conference on Humanoid Robots (Humanoids 2012), pp. 198–203 (2012)

Hybrid Nature-Inspired Optimization Techniques in Face Recognition

Lavika Goel[(⊠)], Abhilash Neog, Ashish Aman, and Arshveer Kaur

Department of Computer Science and Information Systems, BITS Pilani,
Vidya Vihar 333031, Rajasthan, India
lavika.goel@pilani.bits-pilani.ac.in

Abstract. Nature has been a very effective source to develop various Nature Inspired Optimisation algorithms and this has developed into an active area of research. The focus of this paper is to develop a Hybrid Nature-inspired Optimisation Technique and study its application in Face Recognition Problem. Two different hybrid algorithms are proposed in this paper. First proposed algorithm is a hybrid of Gravitational Search Algorithm (GSA) and Big Bang-Big Crunch (BBBC). The other algorithm is an improvement of the first algorithm, which incorporates Stochastic Diffusion Search (SDS) algorithm along with Gravitational Search Algorithm (GSA) and Big Bang-Big Crunch (BB-BC). The hybrid is an enhancement of a single algorithm which when incorporated with similar other algorithms performs better in situations where single algorithms fail to perform well. The algorithm is used to optimize the Eigen vectors generated from Principal Component Analysis. The optimized Eigen faces supplied to SVM classifier provides better face recognition capabilities compared to the traditional PCA vectors. Testing on the face recognition problem, the algorithm showed 95% accuracy in the ORL dataset and better optimization capability on functions like Griewank-rosenbrock, Schaffer F7 in comparison to standard algorithms like Rosenbrock, GA and DASA during the Benchmark Testing.

Keywords: Gravitational Search Algorithm (GSA) · Big Bang-Big Crunch (BB-BC) · Stochastic Diffusion Search (SDS) · Face Recognition Problem · Benchmark Testing

1 Introduction

Optimization is the method of obtaining the most optimal solution(s) for a particular problem. Over the last few decades, with the increase in the complexity of problem, the need and interest in the nature-inspired algorithms has grown strongly. Heuristics are techniques which try to find optimal solutions at minimal cost but without any assurance. Stochastic behavior forms an inseparable part of heuristic algorithms.

In any population based heuristic algorithm, there are two important aspects, exploration and exploitation. The first aspect refers to the ability of expanding the search space in order to find the global optimum whereas the second aspect refers to greedily finding an optimal around a good solution (optimal solution) in the search space. The success

© Springer-Verlag GmbH Germany, part of Springer Nature 2020
M. L. Gavrilova et al. (Eds.): Trans. on Comput. Sci. XXXVI, LNCS 12060, pp. 99–126, 2020.
https://doi.org/10.1007/978-3-662-61364-1_6

of any nature inspired algorithms heavily depends on the precise tuning between exploration and exploitation. To avoid the local minima, in the initial stages of the algorithm exploration aspect is enhanced. As the number of iterations increase, exploration aspect is overshadowed by the exploitation aspect. The key for success of any nature inspired algorithm lies in efficient trade-off between these two aspects.

The application of various nature-inspired techniques like Genetic Algorithm [11], Particle Swarm Optimization [13], Ant-colony optimization [14, 24], etc. can be found in different fields. But despite the success of these algorithms, the fundamental question as to whether there exists any optimization technique that solves all optimisation problems with similar efficiency is answered by No-Free-Lunch theorem (NFL) [12]. According to this theorem, there is no one algorithm to solve all optimisation problems. In other words, an algorithm may be able to solve some problems better and some problem worse than others. Thus, proposal of new high performing methods is always welcome.

Face recognition involves Eigen faces or Eigen vectors whose weighted combination can represent each image of the dataset. The vectors are extracted so as to reduce the dimensionality of the data so that the classifier can be effectively used. The vectors are generated from the covariance matrix which is obtained from the extraction of pixel values. It is very important for the extracted Eigen vectors to have significant contribution in the representation of the original images. So, the problem of face recognition is very much dependent on the Eigen faces obtained.

The aim of this paper is to find better Eigen vectors such that they can replace some of the existing ones, thus enhancing the set of Eigen faces. In this paper, two new hybrid algorithms are proposed which uses the variants of Big Bang-Big Crunch (BB-BC) [5], Gravitational Search Algorithm (GSA) [1] and stochastic diffusion Search (SDS) [7] to create a new approach to get better results. The Eigen faces are considered as the initial points of the search space in the algorithms proposed. The motivation behind the hybridization work is the enhanced performance of a base algorithm in problems like face recognition, when combined with other nature inspired algorithms. BBBC acts as the base algorithm of this hybrid. The intuition behind using BBBC is its great exploration aspect, that takes place during the Big Bang phase, depicting the evolution of universe by random distribution of particles. Drawing similarity from, the evolution of universe and its attainment of a definite order among the objects due to the presence of mass and force of attraction, the points in our search space too, are considered as objects of the universe with certain mass. Thus, the optimization algorithm is defined by imitating the natural movement of particles with mass towards an optimal solution. To define the movement, GSA is applied which considers each particle in the search space as an object with mass, which, by the gravitational force of attraction moves toward each other. Adding GSA to the BBBC enhances the Big Crunch phase. Associating mass and acceleration to each particle allows them to move towards the better solution (heavier particle) in every iteration, which results in a better centre of mass representation during the Big Crunch, leading to a better convergence. SDS acts as a refinement algorithm to the GSA. The movement of particles toward each other (smaller masses towards heavier mass/better solution) based on mass assumes an isolated environment where there is no other force or disturbance. But this is not always the case.

This is where SDS prove useful to GSA. It brings in the factor of external forces on the population. Although the velocity, acceleration and displacement are calculated similarly using GSA, the movement of particles is done based on the SDS hypothesis. Every particle is classified as active or passive based on their displacement from the previous position, using an adaptive threshold. Active diffusion causes the movement of particles in the search space.

The hybrid works better than the algorithms applied singly on the face recognition problem. Although the Big Bang phase of BBBC provide good exploration of the optimal solution, the exploitation is weak, which leads to slow convergence and even suboptimal solution at times. GSA and SDS has low exploration capabilities compared to BBBC, and thus may convergence to a suboptimal solution, although they have a better exploitation phase.

These hybrid algorithms are tested on the ORL Face Recognition dataset and CEC 2015 Benchmark functions.

2 Literature Survey

Algorithms inspired by nature has been effectively applied to face classification problem and has inspired a lot of research since the last decade. Face recognition find its use in various fields like improving the security of biometric systems etc. Eigen faces approach for face recognition was first introduced by Marcus et al. [10] around 1990's. This approach considers a face as a 2D object and a combination of Eigen vectors in an Eigen space, i.e. each image in the database represented as the weighted combination of these Eigen vector (Eigen faces). New faces are similarly written as a weighted sum of these Eigen vectors and assigned to existing class labels by comparing their weights with training set.

Face recognition using labelled graphs was introduced by Wiskott et al. [17] which obtained a graph through a graph matching procedure. Simple Similarity function was used to classify new test images with the known dataset. A new technique of representing faces using binary patterns was introduced in the field of face recognition by Ahonen et al. [18]. Face is divided into multiple parts and each of them is acted upon to extract binary feature distributions which are, in the end, combined to form full face feature descriptor.

Alternative to Eigen faces, the Laplacian faces for face recognition was introduced by He et al. [19] which alike Eigen faces uses locality preserving projections that stores more information about the different parts of the images. There has been many such research carried out on studying face recognition techniques using Eigen Faces [22, 23]. Techniques, like, Principal Component Analysis have enhanced the growth of face recognition techniques using Eigen Faces [26], modular PCA approaches [9], etc.

Number of candidate solutions is a parameter for the general classification of optimization algorithms. Algorithms which start with multiple random solutions and iteratively improves it are called Population-based algorithms. In these algorithms, information can be exchanged among the candidates which helps in finding the optimal solutions but at high computational cost.

Population-based algorithms are better suited to avoid local minima compared to individual-based algorithms. All the population-based algorithms share a common framework which includes two conflicting aspects: exploration vs exploitation. These two aspects are conflicting in nature and promotion of one lead to degradation of the other. The exploration aspect takes care of the stochastic behavior of candidate solution and improves the search space depth. While the exploitation aims to improve the solution quality by searching for solutions locally rather than randomly. Only focusing on the exploration part reduces the chance of getting an optimum solution and only exploitation can lead to local optima and hence a right balance needs to be maintained between the two aspects. Most of the population-based algorithm have a common framework. They all start with a set of random initial solutions and their fitness values are generated using objective functions. Now these candidate solutions are updated such that their fitness value is improved. This process is repeated for multiple iterations till suitable conditions are not met.

Broadly, population-based algorithm can be classified into three different categories, namely evolution, physic, or swarm. Evolutionary algorithms [25] imitate natural evolutionary processes. Biogeography-based Optimization (BBO) algorithm [16], Big Bang Big Crunch (BB-BC) [5] and Asexual Reproduction Optimization (ARO) [15] are the famous evolutionary algorithms. Swarm based algorithms are the ones which are inspired from swarm intelligence. Some of the common swarm-based algorithms are Particle swarm optimization (PSO) [13], Ant Colony Optimization [14, 24], Moth Flame Optimization (MFO) [20] etc. Physics-based algorithms are the ones which imitate any natural phenomenon based on laws of physics. The most common ones are Gravitational Search Algorithm (GSA) [1] and Chemical reaction Optimization (CRO). There are various other meta-heuristic algorithms as well which are bio-inspired. Camilo Caraveo et al. [28] proposed an optimization algorithm based on the self-defense mechanism of plants. Plants have various defense mechanisms to protect itself from insects and other predators. The algorithm is based on this model of predators and preys.

Gravitational search algorithm, a Physics-based optimization was introduced by Rashedi et al. [1] in 2009. This algorithm uses Newton's law of gravity for agents with certain mass to interact among themselves and change their position correspondingly to increase their fitness values. Many modifications have been proposed to further optimize Gravitational Search Algorithm. Gao et al. [2] introduced chaos into GSA to counter problems like local optima entrapment and slow convergence speed. Chaos adds to algorithm randomness and ergodicity to solve these problems. Two variations of the algorithm were proposed. One performs chaotic local search with the current solution as the base solution, in search for better solutions nearby. Other, replaces the random sequences in GSA. Fruman Olivas et al. [27] proposed a method of dynamically adjusting the GSA parameters using type-2 fuzzy logic, which efficiently addresses the issue of tuning the GSA parameters optimally. Rashedi et al. in 2010 [4] proposed a binary version of gravitational search algorithm. In this technique, positions in each dimension are binary in nature. They can take only two values 0 or 1.

Erol et al. [5] first proposed an evolutionary population-based algorithm Big Bang-Big Crunch algorithm. This algorithm has two phases. First phase is Big Bang phase where agents are stochastically scattered in the space. Basically, Energy gets distributed. In the other phase, Big Crunch, agents are brought together using the center of mass concept. Uniform Big-Bang Big-Crunch algorithm was introduced by Alatas et al. [6] where the uniform method was followed to initialize the agent's position and then all agents were crunched to one single point that represents the entire population. This increased convergence as compared to traditional BB-BC. A more robust model of BB-BC, Exponential Big Bang Crunch was proposed by Hasancebi et al. [21]. It was optimized to perform well in code-based design optimization of steel-based structures.

Stochastic diffusion Search, belonging to Swarm intelligence algorithms was first proposed in 1989 by Bishop et al. [7]. This algorithm has two phases, hypothesis test and diffusion. Based on the result of test phase agents are classified as Active agent or Passive agent. During diffusion, active agents share their hypothesis with passive agents leading to better convergence. Information about potentially good solutions are thus spread among all the agents (Al-Rifaie et al. [8]).

3   Existing Methodologies

Algorithms BB-BC, GSA and SDS are hybridized to form a novel nature-inspired optimization algorithm. The Eigen faces [10] generated from Principal Component Analysis are subjected onto this hybrid algorithm to produce an optimized set of Eigen faces [10].

Gravitational Search Algorithm (GSA). This nature-inspired algorithm [1] is based on the Newton's universal law of gravitation, i.e.

$$F = Gm_1m_2/r^2, \text{ where } G = 6.67 \times 10^{-11} \tag{1}$$

Given the mass and force on an object, its acceleration can be calculated as,

$$a = F/m \tag{2}$$

The gravitational constant, G, is defined as a function of time,

$$G(t) = G(t_0) \times (t_0/t)^\beta, \beta < 1 \tag{3}$$

The algorithm considers a multi-dimensional search space with 'N' number of particles. Each particle is defined by its individual mass and position, $X_i = (x_i^1, x_i^2 \ldots x_i^n)$, where $i = 1, 2, \ldots N$ and x_i^d represents the co-ordinate value in the dth dimension. The force experienced by a particle 'i' due to a particle 'j' is,

$$F_{ij}{}^d(t) = G(t) (M_{pi}(t) \times M_{aj}(t))/(R_{ij}(t) + \mathcal{E}) (x_j{}^d(t)\text{-}x_i{}^d(t)) \tag{4}$$

where M_{pi} is the passive gravitational mass of 'i', M_{aj} is the active gravitational mass of j, R(t) is the Euclidean distance between 'i' and 'j', and $\mathcal{E}$ is a constant.

In Eq. (4), the value R was preferred over R^2 as it gave better results experimentally. Applying randomness, the total force on particle 'i' is,

$$F_i^d(t) = \sum_{j=1, \neq i} rand_j F_{ij}^d(t) \qquad (5)$$

Acceleration of particle 'i' is calculated as,

$$a_i^d(t) = F_i^d(t)/M_{ii}(t) \qquad (6)$$

where M_{ii} is the inertial mass of agent 'i'

The position of the ith particle is calculated using acceleration and velocity,

$$v_i^d(t+1) = rand_i \times v_i^d(t) + a_i^d(t)s \qquad (7)$$

$$x_i^d(t+1) = x_i^d(t) + v_i^d(t+1) \qquad (8)$$

The mass M_i of the particle is,

$$m_i(t) = (fit_i - worst(t))/(best(t) - worst(t)) \qquad (9)$$

$$M_i(t) = m_i(t)/\sum_{j=1} m_j(t) \qquad (10)$$

Chaotic Search Algorithm. A search algorithm [3], which act as an enhancer to the traditional gravitational search algorithm. It adds the concept of chaos [2] to the movement of particles governed by GSA. Chaos shows a 'seemingly-random' behavior, which means that, it appears to be random but is actually deterministic. A logistic map can be used to obtain the chaotic behavior. It is given as

$$X_n + 1 = \mu X_n(1 - X_n) \qquad (11)$$

Here, n denotes the iteration number. Chaotic behavior is exhibited when the value of μ is 4. The traditional gravitational search algorithm equations are modified by the introduction of chaotic numbers in place of random numbers.

$$F_i^d(t) = \sum_{j=1, \neq i} chaos_j F_{ij}^d(t) \qquad (12)$$

$$v_i^d(t+1) = chaos_i \times v_i^d(t) + a_i^d(t) \qquad (13)$$

Big Bang Big Crunch (BB-BC). A nature-inspired evolutionary algorithm [5] based on the 'Evolution Theory'. It works in two phases, Big-Bang and Big-Crunch. In the Big-bang phase, agents are randomly distributed in the search space, analogous to Energy being distributed, while in the Big Crunch Phase agents are clustered into a single point using the center of mass concept. The center of mass of the particle population is calculated as,

$$\bar{a}^c = \left(\sum_{i=1}\left(1/f^i\right) \times \bar{a}^i\right)/\left(\sum_{i=1}\left(1/f^i\right)\right) \tag{14}$$

where $\bar{a}^i$ is a point in the search space and f^i is its fitness value.

In the subsequent big-bang phase, population is randomly distributed using normal distribution such that after each iteration the radius of the boundary is decreased, i.e. (the search boundary in the kth iteration)/(Search boundary in the $(k+1)^{th}$ iteration) > 1.

Particles are added using the formula,

$$x_i^{new} = x_c + r \times \alpha \times \left(x_{max} - x_{min}\right)/t \tag{15}$$

where, x_i^{new} is the new agent, α is a parameter used to limit search space and t, the number of iterations.

These two phases continue until the population converge towards an optimal solution.

Stochastic Diffusion Search (SDS). SDS is a population-based Swarm intelligence algorithm [7]. SDS, unlike many nature-inspired algorithms, needs a strong mathematical framework. SDS uses a direct one to one communication between the particles to share a certain hypothesis. The algorithm is dived into two phases: Test Phase and the Diffusion Phase. In the test phase, SDS classifies an agent as Active (good solutions) or passive (bad solutions) based on the hypothesis. In the diffusion phase, information between the Active agent and passive agent is shared through diffusion process [8].

There are three different types of diffusion process, namely active diffusion, passive diffusion and hybrid diffusion. In active, the active agents randomly select an agent from the search space. If the selected agent is passive, hypothesis of the active agent is passed onto it, or shared with the passive agent. In passive diffusion, it is the passive agents that randomly select agents. If an active agent selected, the passive agent starts sharing the active agent's hypothesis, else initialize a new hypothesis. Hybrid diffusion is a mixture of both. The cluster size around an active agent is larger in case of passive diffusion, compared to active diffusion.

4 Proposed Methodology

This section describes the two proposed hybrid algorithms GSA-BBBC with Chaotic Local Search and its modified version, BBBC - GSA - SDS algorithm and their application on face recognition problem. The hybrid algorithms try to mix the best working of all the individual algorithms. BBBC acts as a base algorithm whose second phase gets modified with the addition of GSA and SDS as a preprocessing step before the big crunch. These algorithms are so chosen for the hybridization because they complement each other very well by fitting into the certain phases of the algorithms.

4.1 Hybrid GSA-BBBC with Chaotic Local Search

The proposed algorithm goes through a cycle of four phases. The Big Bang distributes the agents in the search space, which are then allowed to move as per their forces of attraction calculated by the GSA algorithm. Big Crunch merges all the points into a single representative point. A local search is done to check if there exists better solution than the one obtained. The steps are repeated until the algorithm converge towards a stable solution.

Initialization. A modified initialization phase of the Big Bang Big Crunch algorithm is used to initialize the search space. Initially, only two points, P1, P2 are set where P1 = {z1, z2, z3 ... zn} and P2 = {x1, x2, x3, ..., xn}, n is the length of a point. A variable, k, is initialized to 1. We have used a chaotic variable c where $c \neq \{0, 0.25, 0.5, 0.75, 1\}$, instead of the random variable r. The upper and lower limits of the search space, and the maximum population size, is then, determined.

The generation of all points is followed by a random reinitialization of those points found to be defined outside the search space, so that they fall inside the search space boundary. After each iteration, the population size is checked, whether $|P| = |P_s|$, where P_s is the desired population size. The value of k is incremented after the generation of every 2k points. Until the desired population size is achieved, the process is made to continue. Random mass values between 0 and 1 are assigned to each of the generated particles. Acceleration, velocity and force are initialized to zero in all dimensions. The maximum number of iterations for the algorithm to run is determined.

Traversal. Force exerted on each particle by the rest of the population is determined for every particle, using the chaotic force equation (Eq. 12). The acceleration and velocity of each particle is accordingly updated, using Eqs. 6 and 13. Using the updated velocity and previous position, the position vector of each particle is updated.

Big Crunch Phase. The center of mass (the representative point) of all the points in the search space is calculated in this step, using the big crunch formula (Eq. 14) from BB-BC algorithm. This representative point is considered as the best solution for the current iteration and is indicative of where the population may converge to, after multiple iterations. Comparing the current center of mass with that of the previous iteration, if the difference between them is not large, it indicates that the algorithm is converging onto the global best solution. The iterations are halted, if the distance between the current and previous center is less than a previously determined threshold.

Chaotic Local Search. It is an exploitation process to find better solutions than the current best. The search is performed around the current global best solutions, given the high probability of finding similar solution around it. After every iteration, the search radius is decremented, for the algorithm to converge onto a solution. Initialization of a candidate solution takes place in each dimension, and its fitness function value calculated. After every iteration, the best of fitness value among all candidates is calculated. The iterations are halted, if this value is better than the global best's fitness value. Till the search radius gets smaller than a previously determined threshold, the process is repeated. The pseudo code of the algorithm is given below (Fig. 1):

```
1) Population initialized using uniform BB-BC
2) While termination criterion is false, do
   (a) For each candidate
       1. Calculate force using Eq. (12)
       2. Calculate acceleration using Eq. (6)
       3. Update velocity using Eq. (13)
       4. Update position using Eq. (8)
       5. Re-initialize candidate at boundary, if it goes out of search space
       End-for
   (b) Using Big crunch (Eq. 14) calculate the current global best agent
   (c) For all dimensions
       1. Initialize the chaotic variable
       End-for
   (d) While termination criterion not satisfied, do
       1. For each dimension d
          Calculate the fitness value of $X_g^d (k)$
          End-for
       2. Select the local optima $X_g (k)$
       3. Compare and update global best
       4. Search radius update
       5. Candidate's position update
       End-while
   (e) Update candidate masses
End-while
```

Fig. 1. GSA-BBBC pseudo code.

4.2   BB BC-GSA-SDS Algorithm (Modified Approach)

Initialization. To initialize the population, Uniform Big Bang Big Crunch approach is used. This is done to ensure the diversity of initial population. The more diverse the initial candidates are, the more effective the algorithm is. In the beginning only two agents are considered C1 and C2, where $C1 = \{u1, u2, u3 \dots u_n\}$ and $C2 = \{l1, l2, l3, \dots, ln\}$, and k, the dividing factor is 1 and n is the length of the candidate. After the generation of each candidate, they are checked whether they are bounded within the defined search space or not. If not, then that candidate is reinitialized again. After each iteration, 2k candidates are generated, and $|C| = |P|$ is checked where P is the size of desired population. In this way the required population is created and initialized. Each of the generated agent is now considered as a particle (particle agents in GSA) with acceleration and velocity both equal to zero. Each of particle is initially given a random mass value from 0 to 1. The total number of iterations is appropriately determined (can vary).

Traversal. Now for each particle, the force acting on it by the rest of the population is determined using the Force Equation from the GSA technique. Acceleration of each is calculated from the calculated Force and subsequently velocity is calculated from acceleration. Here, the chaotic sequences were used for the acceleration and velocity calculations. Based on the velocity, particle positions are updated.

Diffusion. Now based on the difference in the Updated position and the previous position, (i.e. hypothesis: position of a particle is good if its displacement is less than the threshold), each particle is characterised either as Active or Passive Agent. If the displacement of the particle is above certain threshold value, then it is classified as a Passive Agent else Active agent. This threshold value is dependent on the number of iterations. Initially the threshold value is less, so that more particles can be classified as Passive (Bad hypothesis), so as to ensure the exploration aspect. The threshold value is increased with number of iterations. This ensures that more agents are classified as Active (Good hypothesis) so that exploitation aspect is ensured. In this case threshold value is taken as:

$$\text{Threshold}[i + 1] = \text{Threshold}[i] + \log_{10}\left(i + 1^{\text{th}}\text{iteration}\right) \qquad (16)$$

Now, for each 'active' agent a particle is randomly chosen from the population (Active Searching). If the chosen agent is an active agent then no Diffusion takes places. If it is 'passive', then the 'active agent' shares its position (hypothesis) with the passive one. Another variation to this direct sharing of hypothesis is that the 'active' particle doesn't directly share its hypothesis but forces the particle to move towards itself a certain amount of distance. This distance is a function of the no. of iterations so as to properly tune the exploration and exploitation part.

Big Crunch Phase and Chaotic Local Search. Once the diffusion phase is over, fitness function of the particles is updated. Using the Big crunch formula, the population's center of mass is calculated. This point is considered as the best solution of the iteration as it indicates where the population, after many iterations may converge to. Taking this center of mass (global best) as a base point, a chaotic local search is performed to enhance the exploitation process. This process takes place in multiple iterations and if a solution is found that is better than the current best, then iterations are halted and the obtained solution would be considered as the global best. The search radius is decremented after every iteration making the algorithm converge onto a solution. Now again this process is repeated until the desired termination conditions are met. The pseudo code of this new extended algorithm is given below (Fig. 2):

```
1) Initialize population using uniform BB-BC
2) While termination criterion is false, do
   (a) For each candidate
       1. Calculate force and acceleration using Eq. (12), Eq. (6) respectively
       2. Update velocity and position using Eq. (13), Eq. (8) respectively
       3. Re-initialize candidate at boundary, if it goes out of search space
       4. If Displacement (new-old position)> threshold
              Assign agent "Passive"
          else
              Assign agent "Active"
       End-for
   (b) For all Active agents
       1. Randomly select any agent from set of agents
       2. If selected agent == Active
              continue
          If selected agent == Passive
              Move passive agent to active agent's cluster
       3. Update fitness values
   (c) Using Big crunch (Eq. 14) calculate the current global best agent
   (d) For all dimensions, do
              Initialize chaotic variable
       End-for
   (e) While termination criterion not satisfied do
       1. For all dimension d
              Compute candidate fitness of $X_g{}^d (k)$
              End-for
       2. Select local optima $X_g (k)$
       3. Compare and update global best
       4. Update the Search radius and Candidate's position
       End-while
   (f) Update candidate masses
End-while
```

Fig. 2. GSA-BBBC-SDS pseudo code.

4.2.1 Functional Architecture

The functional architecture of BBBC-GSA-SDS is presented in Fig. 3. The functional architecture mainly comprises four layers BB layer, Gravitational Search layer, SDS layer and BC layer. The explanation of each layer is given below.

A. BBBC Big Bang Layer

Agent Position's Initializer Layer. This layer randomly distributes agents into the search space and hence initializes their positions. This layer outputs Agent's position and feed it to Gravitational search layer

$$X = [x_1, x_2, x_3 \ldots, x_n] \tag{17}$$

The layer initializes the agents in the search space X, which is nothing but a list of all the agents present in the search space.

B. Gravitational Search Layer

Input/Output Layer. This layer takes input Agent's position from Agent Position's Initializer layer and output the updated agent's position caused from the gravitational force.

Force Calculator layer. This layer calculates gravitational force experienced on a particular agent from all other agents in the search space. Force is calculated for all agents using the knowledge of respective positions of all agents.

Acceleration Calculator layer. This layer calculates acceleration of all the agents based on their force and masses. This layer take input from the Force Calculator layer and feeds Agent's acceleration details to Velocity Calculator layer.

Velocity Calculator layer. This layer evaluates velocity of individual agents from their acceleration which was calculated in the previous layer.

Position Calculator layer. Based on the output of Velocity Calculator layer this layer calculates the new position of the agent from its previous position and current velocity.

Boundary checker. This layer checks if the newly calculated position in the previous layer is within the boundaries of search space and gives input to Position layer based on the result. If the newly calculated position is within the boundary then no action is taken, otherwise new position needs to be calculated.

Updated Position Assignment layer. This layer updates the position of the agent after taking input from Position calculator layer and Boundary checker and finally output Agent's position

$$x_i^d(t + 1) = x_i^d(t) + \text{chaos}_i \times v_i^d(t) + a_i^d(t) \tag{18}$$

The equation represents the change in position of an agent due to the forces acting on it. $x_i^d(t + 1)$ is the new position of the agent after its traversal with velocity $\text{chaos}_i \times v_i^d(t)$ and acceleration $a_i^d(t)$.

C. Stochastic Diffusion Layer

Input/Output layer. This layer inputs agent's position from the Gravitational Search Layer and output updated agent's position after the diffusion process.

Displacement Calculator layer. This layer calculates the displacement of each agent based on their updated position and previous position.

Threshold Displacement Comparator. This layer compares the displacement of the agent calculated in the previous Displacement Calculator layer with a particular threshold value and output the result.

Agent Type Assigner. Based on the input from Threshold Displacement Comparator layer, this layer assigns each agent as 'active' or 'passive'. If the displacement is greater than threshold then that agent is assigned 'passive', else assigned 'active'.

Random Agent Selector. This layer randomly selects one agent among all, from the defined search space.

Agent Type checker. This layer checks the type of the agent selected in the Random Agent Selector layer and outputs the result active/passive based on whether it is active/passive agent.

Diffusion layer. Based on the output of Agent Type checker layer, this layer performs diffusion process on the agents. Exchange of hypothesis i.e. moving passive agent towards active agent happens in this layer.

Updated Position Assignment layer. This layer updates the position of the agent after the diffusion process and outputs updated agent's position

$$\left(x_i^d(t+1) - x_i^d(t)\right) < \text{threshold} => \text{Active} = \text{Active} \cup A_i$$
$$\left(x_i^d(t+1) - x_i^d(t)\right) > \text{threshold} => \text{Passive} = \text{Passive} \cup A_i$$

$$\tag{19}$$

$$\forall_k \in \text{Active (if(rand}_j \in \text{Passive)} \; x_j^d(t) = x_k^d(t))$$

The initial part of the equation corresponds to the displacement of an agent, calculated based on the difference between the new and old positions. The displacement is compared with the threshold. If it is less than the threshold, the agent A_i is marked active else passive. The last line of Eq. 19 corresponds to the diffusion process, where, for all active agents k, if a randomly selected agent j is passive, then k's hypothesis is passed onto j, i.e. here, it is the position of active agent k.

D. BBBC Big Crunch Layer

Global Best Agent Calculator. This layer calculates the best agent's position after doing the big crunch operation and using center of mass as global agent's position.

$$\bar{a}^c = \left(\sum_{i=1} \left(1/f^i\right) \times \bar{a}^i\right) / \left(\sum_{i=1} \left(1/f^i\right)\right)$$

$$\tag{20}$$

The equation represents the calculation of center of mass, which basically is a representative point of where the agents may converge in the future iterations. f^i corresponds to the fitness value of the ith agent.

The complete mathematical Framework of BBBC-GSA-SDS Algorithm is represented below:

$$x_i^d(t+1) = x_i^d(t) + chaos_i \times v_i^d(t) + a_i^d(t) \quad \big] \quad \text{GSA Layer}$$

$$(x_i^d(t+1) - x_i^d(t)) > \text{threshold} => \text{Active} = \text{Active} \cup A_i$$
$$(x_i^d(t+1) - x_i^d(t)) > \text{threshold} => \text{Passive} = \text{Passive} \cup A_i \quad \big] \quad \text{SDS Layer}$$
$$\forall \; i \in \text{Active (if (rand}_j \in \text{Passive)} \; x_j^d(t) = x_i^d(t))$$

$$\bar{a}^c = (\sum_{i=1} (1/f^i) \times \bar{a}^i) / (\sum_{i=1} (1/f^i)) \quad \big] \quad \text{Big Crunch Layer}$$

where,

$x_i^d(t + 1)$:   position of ith agent at time $t + 1$ at dimension d
$x_i^d(t)$:   position of ith agent at time t at dimension d
$v_i^d(t)$:   velocity of ith agent at time t at dimension d
$a_i^d(t)$:   acceleration of ith agent at time t at dimension d
threshold: threshold value of active/passive agent
Active: Active agent list
Passive: Passive agent list
A_i: i^{th} agent
$rand_i$:   random no between 0 to 1.
$\bar{a}^c$:   acceleration of center of mass
f^i: fitness value of ith agent

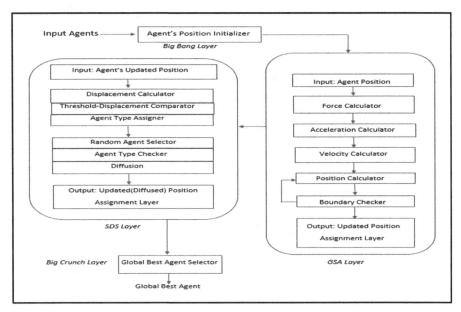

Fig. 3. Functional Architecture of GSA-BBBC and BBBC-GSA-SDS algorithm

5 Face Recognition

There have been many face recognition algorithms developed in the recent times. Out of all, deep neural networks have been the most successful. However, deep neural nets require a large amount of training time and dataset and large computation power, which acts a disadvantage many a times. Our proposed algorithm tries to overcome these difficulties by enhancing the traditional machine learning algorithms. In the application of our algorithm to face recognition, Principal Component Analysis (PCA) technique is used for generating the Eigen faces [10]. On the Eigen faces generated by PCA, the proposed hybrid algorithm is applied. The optimal set of Eigen faces are then used to train an SVM classifier.

BBBC-GSA. The Eigen faces are a representation of the database and each image in the database can be represented as the weighted sum of these Eigen faces. The initial points of the algorithm are the top k Eigen vectors in the Eigen vector space. Each Eigen vector is considered a particle with mass, position, velocity and acceleration. The fitness function of each Eigen vector is the sum of weights of that vector across the training set vectors divided by the total number of vectors. In each iteration, the velocity, acceleration, force and position of every particle is calculated according to the GSA. Based on the forces of attraction, the particles are allowed to move, before being converged onto a single representative point, i.e. the center of mass. With the COM as the base point, a chaotic local search is performed to find better solutions near the current global maxima. The best fit Eigen vector obtained from chaotic local search is compared with the least valued Eigen vector. The solution with the least fitness value is replaced by the one with better fitness value. The steps are repeated until the desired solution is reached.

BBBC-GSA-SDS. The technique is similar to the previous algorithm. Each face in the dataset is represented as a weighted sum of the Eigen vectors. The initial points of the algorithm are the k Eigen vectors in the Eigen vector space. Each Eigen vector is considered as a particle with mass, position, velocity and acceleration. The fitness function is similar to that of BBBC-GSA. In each iteration the velocity, acceleration, force and position of every particle is calculated according to the GSA. Based on the displacement of agents, they are classified as Active or Passive. After the classification, active diffusion takes place between the Active and Passive agents by the sharing of hypothesis. Corresponding fitness values are updated. COM is calculated as per the Big Crunch. With the COM as the base point a chaotic local search is done in order to find better solutions near the current global maxima. The better solution obtained then replaces the Eigen vector with the least fitness value. The steps are repeated until the desired solution is reached.

The flowchart of the face recognition process using BBBC-GSA-SDS is shown in Fig. 4.

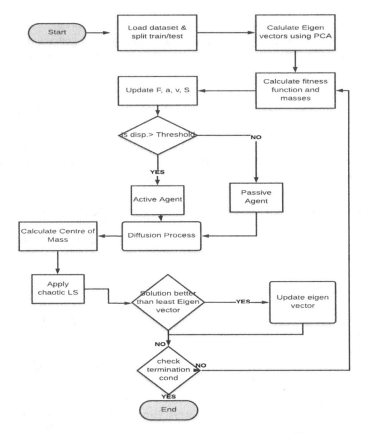

Fig. 4. Flowchart BBBC-GSA-SDS in face recognition

6 Experimental Results

6.1 Benchmark Testing

The hybrid algorithm developed – GSA-SDS-BBBC was tested on CEC 2015 Benchmark functions. The experiments were performed on an Intel Core i7 processor-based system having 8 GB RAM, and running on Windows 10, 64-bit OS. The testing was done using Python 3.6. The CEC-BBOB (Black-box Optimization Benchmarking) 2015 test suite comprises 24 Benchmark functions which are divided into five classes, namely, separable (F1–F5), low or moderate condition (F6–F9), high condition (F10–F14), multi-modal (F15–F19), multi-modal with weak global structure (F20–F24). The algorithm was tested in 2, 3, 5, 10 and 20 dimensions. Number of independent trials per problem instance was set to 1e9. Benchmark Testing done on the 24 standard Benchmark functions, provided considerably comparable results. Some of the good performances was observed in Griewank-rosenbrock F8F2, Rosenbrock rotated, Ellipsoid, Sum of Different powers, Weierstrass, Schaffer F7 condition 10, Schaffer F7 condition 1000, Katsuuras, and Lunacek bi-Rastrigin function.

The algorithm was compared with standard algorithms like Rosenbrock, GA, DASA. The Benchmark Testing results give a good approximation of the comparability of the hybrid algorithm with respect to the standard algorithms. In many cases, the hybrid algorithm performs better than that of Rosenbrock and GA.

The light blue colored curve named hybrid represents the hybrid algorithm. The following graphs shown in Fig. 5 represent the performance of the algorithm on the Griewank-Rosenbrock F8F2 function for dimensions 2, 3, 5, 10 and 20 respectively. The performance of the algorithm on this function improves as dimensions increase. In dimension 20, it is better than all the other compared standard algorithms.

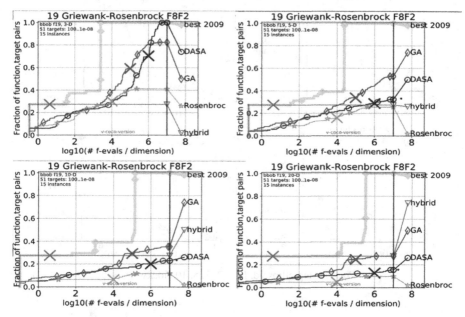

Fig. 5. Griewank-Rosenbrock. Dim-3, 5, 10, 20 clockwise (Color figure online)

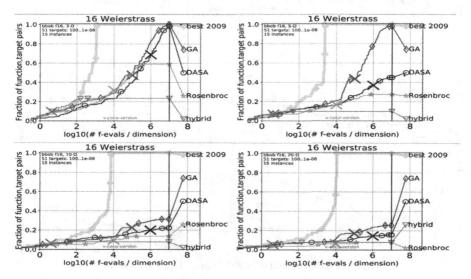

Fig. 6. Weierstrass. Dim-3, 5, 10, 20 clockwise

The figures, Figs. 6, 7 and 8 show the performance of the algorithm on the Weierstrass, Step-ellipsoid, Schaffer F7 condition 1000 function respectively. As the dimensions grow, the results are comparable with that of the standard algorithms. The hybrid algorithm performs better than Rosenbrock on the function Schaffer F7 condition from dimension 10 onwards, and in Weierstrass function in dimension 20.

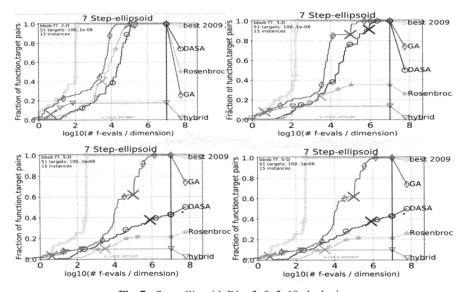

Fig. 7. Step-ellipsoid. Dim-2, 3, 5, 10 clockwise

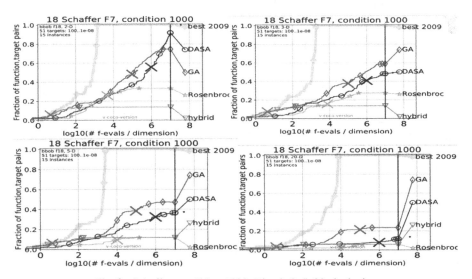

Fig. 8. Schaffer, condition 1000. Dim-2, 3, 5, 20 clockwise

Benchmark Testing on a few more functions have yielded comparable results. The graphs shown below in Figs. 9, 10, 11, 12 and 13 are the results of Benchmark Testing on Lunacek bi-Rastrigin, Schaffer F7 condition 10, Sum of Different Powers, Katsuuras and Rosenbrock rotated function respectively. In Rosenbrock rotated the results are comparable with that of GA as the dimensions grow. The performances of the algorithms in Lunacek bi-Rastrigin are similar from dimension 3 onwards. The hybrid algorithm is on the same line of performance with Rosenbrock algorithm in Schaffer F7 condition function.

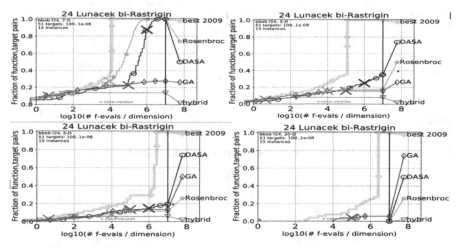

Fig. 9. Lunacek bi-Rastrigin. Dim-2, 3, 5, 20 clockwise

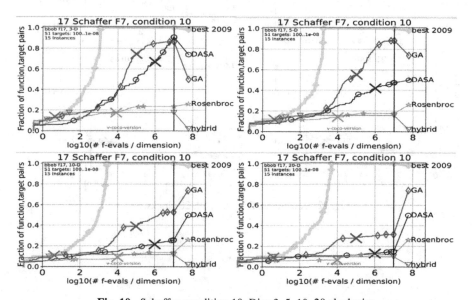

Fig. 10. Schaffer, condition 10. Dim-3, 5, 10, 20 clockwise

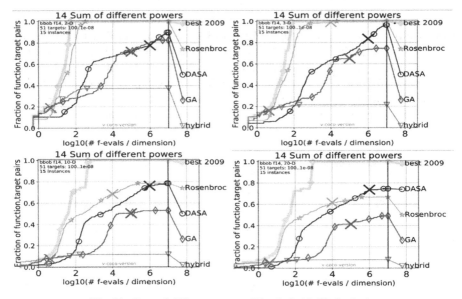

Fig. 11. Sum of different powers. Dim-2, 3, 10, 20 clockwise

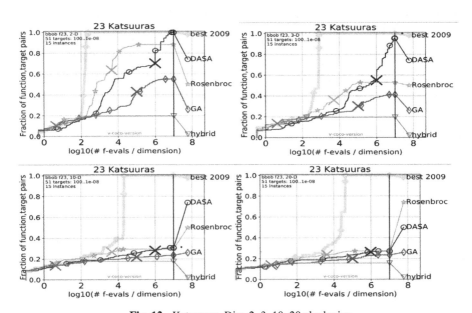

Fig. 12. Katsuuras. Dim-2, 3, 10, 20 clockwise

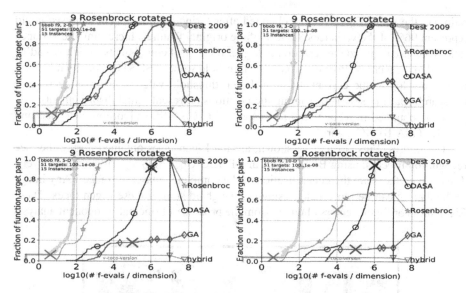

Fig. 13. Rosenbrock. Dim-2, 3, 5, 10 clockwise

The results obtained imply that the proposed algorithm performs well on multi-modal functions having adequate as well as weak global structures, as can be seen from the performance on Griewank-Rosenbrock, Schaffer F7, and Katsuuras. Comparable results are obtained in case of Weierstrass and Lunacek bi-Rastrigin. The algorithm also performs well on functions with high condition, which is evident from the performance graph in Different power function and Schaffer F7 moderately ill-conditioned (condition 100).

6.2 Face Recognition Problem

This proposed methodology was applied to the Face recognition problem. The ORL (Olivetti Research Laboratory) dataset was used during this process. This dataset was specifically chosen because all the images are grayscale images with a homogeneous dark background. And the images contain subjects mostly in a frontal position or with a slight tilt of the head. The consistent gradient of intensity between the subject and background and the consistent subject poses in all the images make the Eigen vectors generated, suitable for optimization by the hybrid algorithm. The size of the dataset is 400 images, in which, there are 10 distinct images for each class (person) and there are 40 distinct classes. Each grayscale image is of the size 92×112 pixels. The maximum number of iterations was taken as one third of the number of Eigen faces. Support vector machines (SVM) was used as the classifier. Repeated experimentations resulted in the parameters of the classifier to be tuned as follows: C value was kept to 1.0, gamma to $1e^{-6}$, and the kernel used was linear. 10% of the data was held out for testing, and the model was trained on the 90% of the data.

Principal Component Analysis was used on the ORL dataset to generate the Eigen faces. These Eigen faces obtained are optimized during the optimization algorithm. The number of principal components taken were 150 as it showed around 96% variance ratio, i.e. the set of Eigen vectors obtained after 'principal component analysis' could represent 96% of the variances, or 96% of the original features. Increasing the variance ratio would've required the selection of higher number of components, which would've resulted in the increase in dimensions. The threshold value in SDS is gradually increased by amount log of the current iteration. Initial threshold value is taken as average of the search space positions (assuming the positions represent initial displacement). The precision obtained in recognizing faces for BBBC-GSA-SDS algorithm was 94% and 95% for GSA-BBBC algorithm.

The figures, Figs. 14 and 15 shown, are the Eigen faces obtained during the face recognition process, using BBBC-GSA and BBBC-GSA-SDS algorithms respectively. These Eigen faces are the generalization of the complete dataset of images. Any image in the dataset can be represented using the weighted sum of these Eigen faces. These Eigen faces or Eigen vectors obtained from the process of Principal component analysis

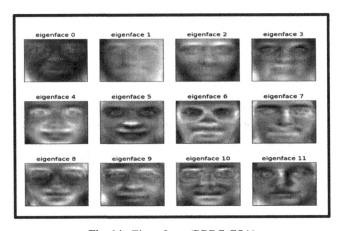

Fig. 14. Eigen faces (BBBC-GSA)

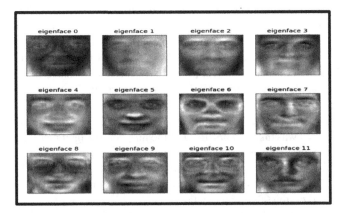

Fig. 15. Eigen faces (BBBC-GSA-SDS)

is basically a low dimension representation of the original high dimensional images, so as to make easier the process of learning.

The tables, Tables 1 and 2 provide the classification reports of the performance of the algorithms in the face recognition problem.

Table 1. Classification report (BBBC-GSA)

	Precision	Recall	F1-score	Support
0	1.00	0.85	0.67	2
1	1.00	1.00	1.00	1
2	0.78	1.00	0.85	1
3	0.87	0.90	0.88	1
4	0.70	0.70	0.70	2
5	1.00	1.00	1.00	2
6	0.98	0.84	0.95	1
7	0.89	1.00	0.95	2
8	1.00	1.00	1.00	1
9	1.00	0.67	0.88	3
10	1.00	1.00	1.00	1
11	0.88	0.90	0.88	1
12	1.00	1.00	1.00	1
13	0.86	0.85	0.85	1
14	1.00	1.00	1.00	1
15	1.00	0.85	0.92	1
16	1.00	1.00	1.00	1
17	0.95	1.00	0.98	1
18	0.90	1.00	0.95	2
19	0.88	1.00	0.95	3
20	1.00	1.00	1.00	2
21	1.00	0.85	0.95	1
22	1.00	1.00	1.00	1
23	1.00	0.90	0.95	1
24	0.92	1.00	0.94	1
25	1.00	1.00	1.00	1
26	1.00	1.00	1.00	1
27	1.00	1.00	1.00	3
Avg/total	0.95	0.94	0.94	40

Table 2. Classification report (BBBC-GSA-SDS)

	Precision	Recall	F1-score	Support
0	1.00	1.00	1.00	2
1	1.00	1.00	1.00	1
2	1.00	0.88	0.94	1
3	0.88	1.00	0.94	1
4	1.00	1.00	1.00	2
5	0.85	0.90	0.87	2
6	0.77	0.88	0.82	1
7	1.00	1.00	1.00	2
8	0.98	1.00	0.99	1
9	0.90	0.90	0.90	3
10	1.00	1.00	1.00	1
11	0.95	1.00	0.98	1
12	1.00	1.00	1.00	1
13	1.00	0.90	0.95	1
14	1.00	1.00	1.00	1
15	1.00	1.00	1.00	1
16	0.80	0.80	0.80	1
17	0.70	0.88	0.78	1
18	1.00	1.00	1.00	2
19	1.00	1.00	1.00	3
20	1.00	1.00	1.00	2
21	0.98	0.90	0.99	1
22	1.00	1.00	1.00	1
23	0.75	1.00	0.86	0
24	0.90	1.00	0.95	1
25	1.00	1.00	1.00	1
26	1.00	1.00	1.00	1
27	1.00	1.00	1.00	3
Avg/total	0.94	0.96	0.95	40

The classification report is comprised of 4 metrics – precision, recall, F1-score and support, based on which the performances are analyzed. Figures 16 and 17 shows the ROC (Receiver Operating Characteristic) curves obtained in case of both the algorithms. It basically represents the performance of the model at all classification thresholds. The curve captures the change in true positive rate (TPR) with respect to the change in the false

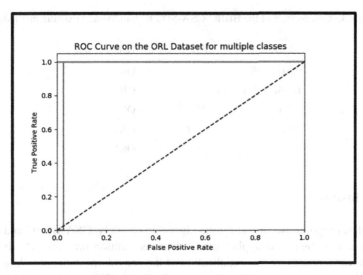

Fig. 16. ROC curve (BBBC-GSA)

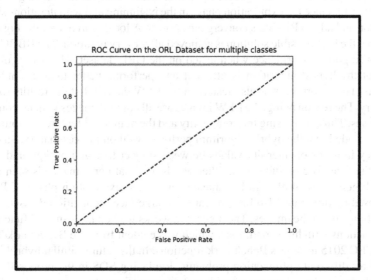

Fig. 17. ROC curve (BBBC-GSA-SDS)

positive rate (FPR). Both, FPR and TPR changes with the change in the classification threshold. The curve captures all the possible instances.

A comparison of our proposed algorithm with the individual algorithms is shown in Table 3.

Table 3. Comparison of the BBBC-GSA-SDS with individual existing methods

Algorithm	F1-Score (%)	Dataset used
BBBC-GSA	94	ORL
BBBC-GSA-SDS	95	ORL
GSA	75–77	ORL
BBBC	60–63	ORL
SDS	91	ORL

7 Conclusion

Our work has proposed two novel hybrid algorithms namely, GSA-BBBC and BBBC-GSA-SDS and studied their application on face recognition problem. The Big-Bang Big-crunch concentrates on the initialization of the candidates in the search space and also controls the convergence of the algorithm. Stochastic Diffusion Search algorithm is used to enhance the convergence power of the Gravitational Search algorithm. And thus, lead to efficient and faster convergence. The parameters chosen for the SDS algorithm are such that they enhance the exploration phase in the beginning and with iterations shifts to exploitation and add to BB-BC's convergence power. A local search after the big crunch ensures that the best possible solution is achieved after every iteration. The BBBC-GSA-SDS showed good performance when tested on the ORL dataset. The results obtained shows that the hybrid algorithm is efficient and performs better than the individual algorithms. The algorithm was also tested on the LFW dataset, but the results were not satisfactory. The reason being, the LFW images are all colored images with no consistent subject poses. Thus, increasing the complexity and the number of features to optimize. It can be concluded that the hybrid algorithm performs well on images that have a certain consistency in terms of intensity values between subject face and background, and in terms of the face pose of subject. In other words, optimal performance is seen on less complex Eigen faces. With a little enhancement in the exploration phase of BB-BC, the proposed method may also be applicable to large face recognition datasets having less variation among the images. The algorithm works well on multi-modal functions as well as functions with high conditions, which was evident from the Benchmark Testing done on CEC 2015 noiseless Benchmark functions. In the future, similar hybrids can be proposed for different optimization problems. Replacing SDS in the proposed hybrid, with Moth-flame algorithm can provide enhanced optimization power as well, as it fits well in the proposed framework. Similar to the face recognition application there are various other problems, like, the travelling salesman problem, where such hybrid techniques can fit into and enhance the working of the existing algorithm.

References

1. Rashedi, E., Nezamabadi-Pour, H., Saryazdi, S.: GSA: a gravitational search algorithm. Inf. Sci. **179**(13), 2232–2248 (2009)

2. Gao, S., Vairappan, C., Wang, Y., Cao, Q., Tang, Z.: Gravitational search algorithm combined with chaos for unconstrained numerical optimization. Appl. Math. Comput. **231**, 48–62 (2014)
3. Choi, C., Lee, J.-J.: Chaotic local search algorithm. Artif. Life Robot. **2**(1), 41–47 (1998). https://doi.org/10.1007/BF02471151
4. Rashedi, E., Nezamabadi-Pour, H., Saryazdi, S.: BGSA: binary gravitational search algorithm. Nat. Comput. **9**(3), 727–745 (2010). https://doi.org/10.1007/s11047-009-9175-3
5. Erol, O.K., Eksin, I.: A new optimization method: Big Bang–Big Crunch. Adv. Eng. Softw. **37**(2), 106–111 (2006)
6. Alatas, B.: Uniform Big Bang–Chaotic Big Crunch optimization. Commun. Nonlinear Sci. Numer. Simul. **16**(9), 3696–3703 (2011)
7. Bishop, J.M.: Stochastic searching networks. In: First IEEE International Conference on Artificial Neural Networks 1989, pp. 329–331. IET (1989)
8. Al-Rifaie, M.M., Bishop, J.M.: Stochastic diffusion search review. Paladyn, J. Behav. Robot. **4**(3), 155–173 (2013)
9. Rajkiran Gottumukkal, V.K.: An improved face recognition techniques based on modular PCA approach. Pattern Recogn. Lett. **25**, 429–436 (2004)
10. Aishwarya, P., Marcus, K.: Face recognition using multiple eigenface subspaces. J. Eng. Technol. Res. **2**, 139–143 (2010)
11. Holland, J.H.: Genetic algorithms. Sci. Am. **267**(1), 66–73 (1992)
12. Wolpert, D.H., Macready, W.G.: No free lunch theorems for optimization. IEEE Trans. Evol. Comput. **1**, 67–82 (1997)
13. Eberhart, R., Kennedy, J.: A new optimizer using particle swarm theory. In: MHS 1995. Proceedings of the Sixth International Symposium on Micro Machine and Human Science, pp. 39–43. IEEE, Nagoya (1995)
14. Colorni, A., Dorigo, M., Maniezzo, V.: Distributed optimization by ant colonies. In: Proceedings of the First European Conference on Artificial Life 1991, pp. 134–142. MIT Press, Cambridge (1991)
15. Farasat, A., Menhaj, M.B., Mansouri, T., Moghadam, M.R.S.: ARO: a new model free optimization algorithm inspired from asexual reproduction. Appl. Soft Comput. **10**, 1284–1292 (2010)
16. Simon, D.: Biogeography-based optimization. IEEE Trans. Evol. Comput. **12**(6), 702–713 (2008)
17. Wiskott, L., Fellous, J.-M., Krüger, N., von der Malsburg, C.: Face recognition by elastic bunch graph matching. In: Sommer, G., Daniilidis, K., Pauli, J. (eds.) CAIP 1997. LNCS, vol. 1296, pp. 456–463. Springer, Heidelberg (1997). https://doi.org/10.1007/3-540-63460-6_150
18. Ahonen, T., Hadid, A., Pietikainen, M.: Face description with local binary patterns: application to face recognition. IEEE Trans. Pattern Anal. Mach. Intell. **12**, 2037–2041 (2006)
19. He, X., Yan, S., Hu, Y., Niyogi, P., Zhang, H.J.: Face recognition using Laplacian faces. IEEE Trans. Pattern Anal. Mach. Intell. **3**, 328–340 (2005)
20. Mirjalili, S.: Moth-flame optimization algorithm: a novel nature-inspired heuristic paradigm. Knowl. Based Syst. **89**, 228–249 (2015)
21. Hasançebi, O., Azad, S.K.: An exponential big bang-big crunch algorithm for discrete design optimization of steel frames. Comput. Struct. **110**, 167–179 (2012)
22. Turk, M., Pentland, A.: Eigenfaces for recognition. J. Cogn. Neurosci. **3**(1), 71–86 (1991)
23. Ravi, S., Nayeem, S.: A study on face recognition technique based on Eigenface. Int. J. Appl. Inf. Syst. **5**(4), 57–62 (2013)
24. Dorigo, M., Blum, C.: Ant colony optimization theory: a survey. Theoret. Comput. Sci. **344**(2–3), 243–278 (2005)
25. Preux, P., Talbi, E.G.: Towards hybrid evolutionary algorithms. Int. Trans. Oper. Res. **6**(6), 557–570 (1999)

26. Al-Arashi, W.H., Ibrahim, H., Suandi, S.A.: Optimizing principal component analysis performance for face recognition using genetic algorithm. Neurocomputing **128**, 415–420 (2014)
27. Olivas, F., Valdez, F., Melin, P., Sombra, A., Castillo, O.: Interval type-2 fuzzy logic for dynamic parameter adaptation in a modified gravitational search algorithm. Inf. Sci. **476**, 159–175 (2019)
28. Caraveo, C., Valdez, F., Castillo, O.: A new optimization meta-heuristic algorithm based on self-defense mechanism of the plants with three reproduction operators. Soft. Comput. **22**(15), 4907–4920 (2018). https://doi.org/10.1007/s00500-018-3188-8

Author Index

Printed in the United States
By Bookmasters